Springer Tracts in Modern Physics
Volume 172

Managing Editor: G. Höhler, Karlsruhe

Editors: J. Kühn, Karlsruhe
Th. Müller, Karlsruhe
A. Ruckenstein, New Jersey
F. Steiner, Ulm
J. Trümper, Garching
P. Wölfle, Karlsruhe

Honorary Editor: E. A. Niekisch, Jülich

Now also Available Online

Starting with Volume 163, Springer Tracts in Modern Physics is part of the Springer LINK service. For all customers with standing orders for Springer Tracts in Modern Physics we offer the full text in electronic form via LINK free of charge. Please contact your librarian who can receive a password for free access to the full articles by registration at:

http://link.springer.de/series/stmp/reg_form.htm

If you do not have a standing order you can nevertheless browse through the table of contents of the volumes and the abstracts of each article at:

http://link.springer.de/series/stmp/

There you will also find more information about the series.

Springer
Berlin
Heidelberg
New York
Barcelona
Hong Kong
London
Milan
Paris
Singapore
Tokyo

Springer Tracts in Modern Physics

Springer Tracts in Modern Physics provides comprehensive and critical reviews of topics of current interest in physics. The following fields are emphasized: elementary particle physics, solid-state physics, complex systems, and fundamental astrophysics.
Suitable reviews of other fields can also be accepted. The editors encourage prospective authors to correspond with them in advance of submitting an article. For reviews of topics belonging to the above mentioned fields, they should address the responsible editor, otherwise the managing editor.
See also http://www.springer.de/phys/books/stmp.html

Managing Editor

Gerhard Höhler
Institut für Theoretische Teilchenphysik
Universität Karlsruhe
Postfach 69 80
76128 Karlsruhe, Germany
Phone: +49 (7 21) 6 08 33 75
Fax: +49 (7 21) 37 07 26
Email: gerhard.hoehler@physik.uni-karlsruhe.de
http://www-ttp.physik.uni-karlsruhe.de/

Elementary Particle Physics, Editors

Johann H. Kühn
Institut für Theoretische Teilchenphysik
Universität Karlsruhe
Postfach 69 80
76128 Karlsruhe, Germany
Phone: +49 (7 21) 6 08 33 72
Fax: +49 (7 21) 37 07 26
Email: johann.kuehn@physik.uni-karlsruhe.de
http://www-ttp.physik.uni-karlsruhe.de/~jk

Thomas Müller
Institut für Experimentelle Kernphysik
Fakultät für Physik
Universität Karlsruhe
Postfach 69 80
76128 Karlsruhe, Germany
Phone: +49 (7 21) 6 08 35 24
Fax: +49 (7 21) 6 07 26 21
Email: thomas.muller@physik.uni-karlsruhe.de
http://www-ekp.physik.uni-karlsruhe.de

Fundamental Astrophysics, Editor

Joachim Trümper
Max-Planck-Institut für Extraterrestrische Physik
Postfach 16 03
85740 Garching, Germany
Phone: +49 (89) 32 99 35 59
Fax: +49 (89) 32 99 35 69
Email: jtrumper@mpe-garching.mpg.de
http://www.mpe-garching.mpg.de/index.html

Solid-State Physics, Editors

Andrei Ruckenstein
Editor for The Americas
Department of Physics and Astronomy
Rutgers, The State University of New Jersey
136 Frelinghuysen Road
Piscataway, NJ 08854-8019, USA
Phone: +1 (732) 445 43 29
Fax: +1 (732) 445-43 43
Email: andreir@physics.rutgers.edu
http://www.physics.rutgers.edu/people/pips/Ruckenstein.html

Peter Wölfle
Institut für Theorie der Kondensierten Materie
Universität Karlsruhe
Postfach 69 80
76128 Karlsruhe, Germany
Phone: +49 (7 21) 6 08 35 90
Fax: +49 (7 21) 69 81 50
Email: woelfle@tkm.physik.uni-karlsruhe.de
http://www-tkm.physik.uni-karlsruhe.de

Complex Systems, Editor

Frank Steiner
Abteilung Theoretische Physik
Universität Ulm
Albert-Einstein-Allee 11
89069 Ulm, Germany
Phone: +49 (7 31) 5 02 29 10
Fax: +49 (7 31) 5 02 29 24
Email: steiner@physik.uni-ulm.de
http://www.physik.uni-ulm.de/theo/theophys.html

Daniel Braun

Dissipative Quantum Chaos and Decoherence

With 22 Figures

 Springer

Dr. Daniel Braun
University of Essen
Institute of Physics
Universitätsstrasse 5
45117 Essen, GERMANY
E-mail: daniel@indy1.theo-phys.uni-essen.de

Library of Congress Cataloging-in-Publication Data.
Die Deutsche Bibliothek - CIP-Einheitsaufnahme
Braun, Daniel:
Dissipative quantum chaos and decoherence/Daniel Braun. - Berlin;
Heidelberg ; New York ; Barcelona ; Hong Kong ; London ; Milan ; Paris;
Singapore; Tokyo: Springer, 2001
(Springer tracts in modern physics; Vol. 172)
(Physics and astronomy online library)
ISBN 3-540-41197-6

Physics and Astronomy Classification Scheme (PACS): 03.65.Sq, 03.67.Lx, 05.20.Gg

ISSN print edition: 0081-3869
ISSN electronic edition: 1615-0430
ISBN 3-540-41197-6 Springer-Verlag Berlin Heidelberg New York

This work is subject to copyright. All rights are reserved, whether the whole or part of the material is concerned, specifically the rights of translation, reprinting, reuse of illustrations, recitation, broadcasting, reproduction on microfilm or in any other way, and storage in data banks. Duplication of this publication or parts thereof is permitted only under the provisions of the German Copyright Law of September 9, 1965, in its current version, and permission for use must always be obtained from Springer-Verlag. Violations are liable for prosecution under the German Copyright Law.

Springer-Verlag Berlin Heidelberg New York
a member of BertelsmannSpringer Science+Business Media GmbH

© Springer-Verlag Berlin Heidelberg 2001
Printed in Germany

The use of general descriptive names, registered names, trademarks, etc. in this publication does not imply, even in the absence of a specific statement, that such names are exempt from the relevant protective laws and regulations and therefore free for general use.

Typesetting: Camera-ready copy from the author using a Springer LaTeX macro package
Cover design: *design & production* GmbH, Heidelberg

Printed on acid-free paper   SPIN: 10777023   56/3141/tr   5 4 3 2 1 0

Preface

One hundred years after the discovery of the first foundations of quantum mechanics, there are still many open and fascinating questions dealing with the relation between quantum mechanics and classical mechanics. Everyday life and three and a half centuries of successful application of classical mechanics have left us with the conviction that we can predict precisely the fate of an individual object if we know sufficiently precisely its initial conditions and the forces that act on it. Quantum mechanics gives us a very different picture of reality. It states that the information that we may gather about any object can never be as complete as in classical mechanics, and we can only predict statistical distributions for experimental data.

Shortly before the beginning of the twentieth century, Henri Poincaré discovered that even within classical mechanics the predictability of very simple classical systems might be very poor, and for sufficiently long times prediction might be just impossible owing to a very strong sensitivity to initial conditions. Such systems were later termed "chaotic". We know today that chaotic behavior is far more common in nature than the regular, integrable motion in, say, Kepler's problem or the harmonic oscillator. It is therefore natural to abandon the attempt to predict the fate of individual objects, for initial conditions are never precisely known. By going over to an ensemble description, as is also done in statistical mechanics, one allows space for uncertainties in the initial conditions. Furthermore, within an ensemble description classical mechanics uses a vocabulary that is much more similar to that of quantum mechanics. Both then predict an evolution of the probability distributions of observables, and we can study how the quantum mechanical evolution law goes over into the classical one.

Nevertheless, the transition from quantum mechanics to classical mechanics is still far from simple, as the classical limit is highly singular. An initial "blob" corresponding to a reasonably localized distribution in phase space is rapidly torn apart by a chaotic classical dynamics, which stretches and folds it to ever finer scales while covering rapidly the entire available phase space. Heisenberg's uncertainty relation, on the other hand, prevents the production of arbitrarily fine scales by quantum mechanical time evolution. And yet another difference between the quantum mechanical world and the classical world exists: probabilities add very differently in the cases of quantum mechanics and classical mechanics. In quantum mechanics probability amplitudes that are squared to give probability distributions have to be added, and this can give rise to quantum mechanical interference effects. In classical mechanics we add probabilities directly and quantum mechanical interference is absent.

It has become obvious during the last twenty years that an important ingredient of the transition from quantum mechanics to classical mechanics is the interaction of a system with its environment. Such a coupling leads typically to dissipation of energy and to decoherence. While the former process is already present in classical mechanics, and by itself leads to a washing out of phase space structures (although on classical scales), decoherence is genuinely quantum mechanical and means that interference patterns are destroyed. Thus, decoherence is the process that allows us to recover classical probability theory from the quantum mechanical theory.

The relations and connections of the quantum mechanical time evolution to the classical evolution for systems that are coupled to an environment are the main subject of this book. The book deals mostly with systems with large quantum numbers, i.e. a semiclassical regime. A new formalism is developed that allows us to efficiently calculate the effects of dissipation and decoherence. It turns out that many of the concepts, such as periodic-orbit theory, trace formulae and zeta functions, that have been introduced to deal with the quantum mechanics of classically chaotic but isolated systems can be extended to situations where dissipation and decoherence are important. Furthermore, I shall deal in some detail with exceptional situations where decoherence is very weak in spite of a strong coupling to the environment. In the young theory of quantum computing, such situations have gained substantial interest in the last few years.

The present book would not have been possible without the help and support of many people. It is my pleasure to thank Prof. Fritz Haake for giving me the opportunity to work on this project in Essen and for his continuous interest, countless discussions and ideas. His enthusiasm and his warm and encouraging support made it a pleasure to work with him.

I would also like to thank Prof. Petr A. Braun, with whom I had the privilege to work closely. With pleasure I think back to his visits to Essen, and to his warm hospitality during my stay in St. Petersburg.

A big "thank you" also to Profs. Marek Kuś and Karol Życzkowksi, frequent visitors to Essen, with whom I have enjoyed working.

During my time in Essen and at numerous conferences and workshops and on visits, I had the pleasure to meet and have discussions with many physicists. Special thanks are owed to Alex Altland, Tobias Brandes, Andreas Buchleitner, Doron Cohen, Predrag Cvitanovic, David DiVincenzo, Bruno Eckhardt, Klaus Frahm, Yan Fyodorov, Pierre Gaspard, Theo Geisel, Nicolas Gisin, Sven Gnutzmann, Martin Gutzwiller, Peter Hänggi, Serge Haroche, Etienne Hofstetter, Martin Janssen, Maria José-Sanchez, Stefan Kettemann,

Roland Ketzmerick, Ilki Kim, Peter Knight, Bernhard Kramer, Wolfgang Lange, Angus MacKinnon, Günter Mahler, Gilles Montambaux, Michael Pascaud, Frédéric Piéchon, Sarben Sarkar, Rüdiger Schack, Ferdinand Schmidt-Kaler, Henning Schomerus, Petr Šeba, Pragya Shukla, Uzy Smilansky, Hans-Jürgen Sommers, Andrew Steane, Frank Steiner, Walter Strunz, Mikhail Titov, Imre Varga, Gábor Vattay, David Vitali, Jürgen Vollmer, Joachim Weber, Ulrich Weiß, Christoph Wunderlich, Hugo Zbinden, Isa Zharekeshev and Wojciech Zurek for the enrichment they brought to my knowledge of physics relevant to the topics of this book.

The numerical calculations were partly performed at the John von Neumann Center for Computing (the former Hochleistungsrechenzentrum Jülich) in Jülich. This work was supported by the Sonderforschungsbereich 237 "Unordnung und große Fluktuationen" (DFG special research program 237, "Disorder and large fluctuations").

Essen, October 2000   *Daniel Braun*

Contents

1. **Introduction** 1
2. **Classical Maps** 7
   2.1 Definition and Examples 7
   2.2 Classical Chaos 9
   2.3 Ensemble Description 11
       2.3.1 The Frobenius–Perron Propagator 11
       2.3.2 Different Types of Classical Maps 12
       2.3.3 Ergodic Measure 15
       2.3.4 Unitarity of Classical Dynamics 16
       2.3.5 Spectral Properties of the Frobenius–Perron Operator . 17
   2.4 Summary 18

3. **Unitary Quantum Maps** 21
   3.1 What is a Unitary Quantum Map? 21
   3.2 A Kicked Top 22
   3.3 Quantum Chaos for Unitary Maps 24
   3.4 Semiclassical Treatment of Quantum Maps 27
       3.4.1 The Van Vleck Propagator 27
       3.4.2 Gutzwiller's Trace Formula 28
   3.5 Summary 29

4. **Dissipation in Quantum Mechanics** 31
   4.1 Generalities 31
   4.2 Superradiance Damping in Quantum Optics 33
       4.2.1 The Physics of Superradiance 33
       4.2.2 Modeling Superradiance 34
       4.2.3 Classical Behavior 36
   4.3 The Short-Time Propagator 37
   4.4 The Semiclassical Propagator 40
       4.4.1 Finite-Difference Equation 40
       4.4.2 WKB Ansatz 40
       4.4.3 Hamiltonian Dynamics 41
       4.4.4 Solution of the Hamilton–Jacobi Equation 42

        4.4.5 WKB Prefactor 43
        4.4.6 The Dissipative Van Vleck Propagator 44
        4.4.7 Propagation of Coherences 45
        4.4.8 General Properties of the Action R 47
        4.4.9 Numerical Verification 47
        4.4.10 Limitations of the Approach 48
    4.5 Summary... 49

5. Decoherence 51
    5.1 What is Decoherence?..................................... 51
    5.2 Symmetry and Longevity: Decoherence-Free Subspaces 53
    5.3 Decoherence in Superradiance 55
        5.3.1 Angular-Momentum Coherent States................ 55
        5.3.2 Schrödinger Cat States 56
        5.3.3 Initial Decoherence Rate 56
        5.3.4 Antipodal Cat States 57
        5.3.5 General Result at Finite Times..................... 57
        5.3.6 Preparation and Measurement 58
        5.3.7 General Decoherence-Free Subspaces 60
    5.4 Summary... 62

6. Dissipative Quantum Maps 63
    6.1 Definition and General Properties 63
        6.1.1 Type of Maps Considered 65
    6.2 A Dissipative Kicked Top 65
        6.2.1 Classical Behavior 66
        6.2.2 Quantum Mechanical Behavior 68
    6.3 Ginibre's Ensemble 71
    6.4 Summary... 73

7. Semiclassical Analysis of Dissipative Quantum Maps 75
    7.1 Semiclassical Approximation for the Total Propagator 75
    7.2 Spectral Properties 78
        7.2.1 The Trace Formula 78
        7.2.2 Numerical Verification 85
        7.2.3 Leading Eigenvalues 88
        7.2.4 Comparison with RMT Predictions 95
    7.3 The Wigner Function and its Propagator 100
    7.4 Consequences 106
        7.4.1 The Trace Formula Revisited 106
        7.4.2 The Invariant State............................. 106
        7.4.3 Expectation Values 108
        7.4.4 Correlation Functions 108
    7.5 Trace Formulae for Expectation Values
        and Correlation Functions 111

	7.5.1 The General Strategy 111
	7.5.2 Cycle Expansion 112
	7.5.3 Newton Formulae for Expectation Values 114
7.6	Summary... 116

**A. Saddle-Point Method
for a Complex Function
of Several Arguments** 119

**B. The Determinant of a Tridiagonal,
Periodically Continued Matrix** 121

**C. Partial Classical Maps
and Stability Matrices
for the Dissipative Kicked Top** 123
   C.1 Rotation by an Angle β About the y Axis 123
   C.2 Torsion About the z Axis................................ 124
   C.3 Dissipation 124

References 125

Index 131

1. Introduction

The notion of "chaos" emerged in classical physics about a century ago with the pioneering work of Poincaré. After two and a half centuries of application of Newton's laws to more and more complicated astronomical problems, he was privileged to discover that even in very simple systems extremely complicated and unstable forms of motion are possible [1]. It seems that this first appeared a curiosity to his contemporaries. Moreover, quantum mechanics and relativistic mechanics were soon to be discovered and distracted most of the attention from classical problems. In any case, classical chaos interested mostly only mathematicians, from G. Birkhoff in the 1920s to Kolmogorov and his coworkers in the 1950s. Only Einstein, as early as 1917, i.e. even before Schrödinger's equation was invented, clearly saw that chaos in classical mechanics also posed a problem in quantum mechanics [2]. The rest of the world started to realize the importance of chaos only when computers allowed us to simulate simple physical systems. It then became obvious that integrable systems, with their predictable dynamics, that had been the backbone of physics for by then three centuries were an exception. Almost always there are at least some regions in phase space where the dynamics becomes irregular and very sensitive to the slightest changes in the initial conditions. The in principle perfect predictability of classical systems over arbitrary time intervals given a precise knowledge of all initial positions and momenta of all particles involved is entirely useless for such "chaotic" systems, as initial conditions are *never* precisely known.

The understanding of quantum mechanics naturally developed first of all with the solution of the same integrable systems known from classical mechanics, such as the hydrogen atom (as a variant of Kepler's problem) or the harmonic oscillator. With the growing conviction that integrable systems are a rare exception, it became natural to ask how the quantum mechanical behavior of systems whose classical counterpart is chaotic might look. Research in this direction was pioneered by Gutzwiller. In the early 1970s he published a "trace formula" which allows one to calculate the spectral density of chaotic systems [3, 4]. That work was extended later by various researchers to other quantities, such as transition matrix elements and correlation functions of observables. All of these theories are "semiclassical" theories. They make use of classical information, in particular classical periodic orbits, their actions and their stabilities, in order to express quantum mechanical quantities. And they are (usually first-order) asymptotic expansions in $\hbar$ divided by a typical action.

The true era of quantum chaos started, however, with the discovery by Bohigas and Giannoni [5] and Berry [6] and their coworkers in the early 1980s that the quantum energy spectra of classically chaotic systems show universal spectral correlations, namely correlations that are described by random-matrix theory (RMT). The latter theory, developed by Wigner, Dyson, Mehta and others starting from the 1950s, assumes that the Hamilton operator of a complex system can be well represented by a random-matrix restricted only by general symmetry requirements. Since there are no physical parameters in the theory (other than the mean level density, which, however, has to be rescaled to unity for any physical system before it can be compared with RMT), the predicted spectral correlations are completely universal. Over the years, overwhelming experimental and numerical evidence has been accumulated for this so called "random-matrix conjecture" – but still no definitive proof is known.

With the help of Gutzwiller's semiclassical theory, Berry has shown that the spectral form factor (i.e. the Fourier transform of the autocorrelation function of spectral density fluctuations) should agree with the RMT prediction, at least for small times [7]. How small these times should be is arguable, but at most they can be the so-called Heisenberg time, $\hbar$ divided by the mean level spacing at the relevant energy. From the derivation itself, one would expect a much earlier breakdown, namely after the "Ehrenfest time" of order $h^{-1}\ln\hbar_{\text{eff}}$, in which h means the Lyapunov exponent and $\hbar_{\text{eff}}$ an "effective" $\hbar$. At that time the average distance between periodic orbits becomes so small that the saddle-point approximation underlying Gutzwiller's trace formula is expected to become unreliable.

In his derivation Berry uses a "diagonal approximation" which is effectively a classical approximation: the fluctuations of the density of states are expressed by Gutzwiller's trace formula as a sum over periodic orbits. Each orbit contributes a complex number with a phase given by the action of the orbit in units of $\hbar$. In the spectral form factor the product of two such sums enters, and in the diagonal approximation only the "diagonal" terms are kept, with the result that the corresponding phases cancel. The off-diagonal terms are assumed to vanish if an average over a small energy window is taken, since they oscillate rapidly. For times larger than the Heisenberg time the off-diagonal terms cannot be neglected, and so far it has only been possible to extract the long-time behavior of the form factor approximately and with additional assumptions by bootstrap methods that use the unitarity of the time evolution, relating the long-time behavior to the short-time behavior [8].

The question arose as to whether semiclassical methods might work better if a small amount of dissipation was present. Dissipation of energy introduces, almost unavoidably, decoherence, i.e. it destroys quantum mechanical interference effects. Therefore dissipative systems are expected to behave more classically from the very beginning, and so one might indeed expect an improvement. To answer this question was a main motivation for the present work. As for most simple questions, the answer is not simple, though: in some aspects the semiclassical theories do work better, in others they do not.

First of all, there are aspects of the semiclassical theory that seem to work as well with dissipation as without. One of them is the existence of a Van Vleck propagator, an approximation of the exact quantum propagator to first order in the effective $\hbar$. Gutzwiller's theory is based on it in the case without dissipation. And a corresponding semiclassical approximation can be obtained for a pure relaxation process by means of the well-known WKB approximation.

Things become more complicated because of the fact that a density matrix, not a wave function, should be propagated if dissipation of energy is included (alternatively, one might resort to a quantum state diffusion approach, as was done numerically in [9], but then one has to average over many runs). If the wave function lives in a d-dimensional Hilbert space, the density matrix has d^2 elements, and its propagator P is a $d^2 \times d^2$ matrix, instead of a $d \times d$ matrix as for the propagator F of the wave function. This implies that many more traces (i.e. traces of powers of P) are needed if one wants to calculate all the eigenvalues of P.

Furthermore, the eigenvalues of P move into the unit circle when dissipation is turned on. For arbitrary small dissipation and small enough effective $\hbar$ their density increases exponentially towards the center of the unit circle. This has the unpleasant consequence that numerical routines that reliably recover eigenvalues of F *on* the unit circle from the traces of F become highly unstable. They fail even for rather modest dimensions, even if the numerically "exact" traces are supplied – not to mention semiclassically calculated ones that are approximated to lowest order in the effective $\hbar$. This must be contrasted with the case of energy-conserving systems, where it has been possible to calculate very many energy levels, e.g. for the helium atom [10] or for hydrogen in strong external electric and magnetic fields [11, 12], or even entire spectra for small Hilbert space dimensions [13].

But dissipation of energy does improve the status of semiclassical theories in various other respects. First of all, the diagonal approximation, which is not very well controlled for unitary time evolutions, can be rigorously *derived* if a small amount of dissipation is present. As a result one obtains an entirely *classical* trace formula, namely the traces of the Frobenius–Perron operator that propagates phase space density for the corresponding classical system. Periodic orbits of a *dissipative* classical map are now the decisive ingredients, and there is a much richer zoo of them compared with nondissipative systems. Fixed points can now be point attractors or repellers, and the overall phase space structure is usually a strange attractor. The traces are entirely real, and no problems with rapidly oscillating terms arise, nor are Maslov indices needed. The absence of the latter in the classical trace formula cannot be appreciated enough, as their calculation can in practice be rather difficult. The ignorance of the Maslov phases seems to have prevented, for example, a semiclassical solution of the helium atom for more than 70 years, in spite of heroic efforts by many of the founding fathers of quantum mechanics before this was done correctly by Wintgen et al. [10] (see the historical remarks in [14]).

Despite the numerical difficulties in the calculation of eigenvalues, the semiclassically obtained traces can be used to reliably obtain the *leading* eigenvalues, i.e. the eigenvalues with the largest absolute values of the quantum mechanical propagator, from just a few classical periodic orbits. These eigenvalues become independent of the effective $\hbar$ if the latter is small enough, and they converge to the leading complex eigenvalues of the Frobenius–Perron operator P_{cl}, the so-called Ruelle resonances. All time-dependent expectation values and correlation functions carry the signature of these resonances, as well as the decaying traces of P themselves. So a little bit of dissipation (an "amount" that vanishes in the classical limit is enough, as we shall see) ensures that the *classical* Ruelle resonances determine the *quantum mechanical* behavior.

As for the range of validity of the semiclassical results, there seems to be no improvement at first glance. The trace formula for the dissipative system is valid at most up to the Heisenberg time of the dissipation-free system, but is eventually limited to the Ehrenfest time for the same technical reasons as for the periodic-orbit theory for nondissipative systems. But this *is* in fact an enormous improvement: for small values of the effective $\hbar$ all correlation functions, traces etc. have long ago decayed to their stationary values before the Heisenberg time (which typically increases with decreasing effective $\hbar$) or, for exponentially small effective $\hbar$, even before the Ehrenfest time is reached, just because the decay happens on the classical and therefore $\hbar$-independent time-scales set by the Ruelle resonances. Only exponentially small corrections to the stationary value are left at the Heisenberg time. One may therefore say that the semiclassical analysis is valid over the entire *relevant* time regime – something one cannot so easily claim for unitary time evolutions.

The important aspect of dissipation that makes quantum mechanical systems look more classical is not dissipation of energy itself, but decoherence. It was long believed that decoherence is an inevitable fact if a system couples to its environment. In particular, it typically restricts the existence of superpositions of macroscopically distinct states, so-called Schrödinger cats, to extremely small times. That is one of the main reasons why these beasts are never observed! However, in the course of our investigations of dissipative quantum maps we have found that exceptions are possible. If the system couples to the environment in such a way that different states acquire exactly the same time-dependent phase factor owing to a symmetry in the coupling to the environment, those states will remain phase coherent, regardless of how macroscopically distinct they are. Similar conclusions were drawn at the same time in the young field of quantum computing. Decoherence is *the* main obstacle to actual implementations of quantum computers. An entire chapter in this book is therefore devoted to the decoherence question. I investigate, in particular, implications for a system of N two-level atoms in a cavity that has potential interest for quantum computing. It turns out that a huge decoherence-free subspace in Hilbert space exists, whose dimension grows exponentially with the number of atoms.

The present book is intended to be sufficiently self-contained to be understandable to a broad audience of physicists. The main parts are concerned with dissipative quantum maps. Maps arise in a natural way mostly from periodically driven systems, and have many advantages (discussed in detail in Chap. 2) that make them favorable compared with autonomous systems. Experts familiar with classical maps, Frobenius–Perron operators and quantum maps may skip Chaps. 2 and 3, which introduce these concepts.

In Chap. 4 I derive the semiclassical propagator for a relaxation process that will underly all of the subsequent semiclassical analysis. The derivation closely follows the original derivation published in [15], but the importance of this propagator and the desire to make the presentation self-contained justify including the derivation once more in the present book. Chapter 5 deals in detail with decoherence, and Chap. 6 presents an overview of the known properties of a dissipative kicked top that will serve as a model system for the rest of the book. Most of the semiclassical results are contained in the long Chap. 7, in particular the derivation of the trace formula, the extraction of the leading eigenvalues, and the calculation of time-dependent observables and correlation functions.

2. Classical Maps

Let us warm up with a brief introduction to classical chaos in the context of classical maps. I shall first define what I mean by a classical map and present a few examples. A precise definition of classical chaos will follow, and I shall emphasize in particular some implications for dissipative maps, which are the main topic of this book. An ensemble description of the classical dynamics will lead to the introduction of the Frobenius–Perron propagator of the phase space density. This operator will also play an important role later on in the context of dissipative quantum maps, since it will turn out that many properties of the quantum propagators are related to the corresponding properties of the Frobenius–Perron propagator.

2.1 Definition and Examples

A classical map $\boldsymbol{f}_{\mathrm{cl}}$ is a map of phase space onto itself. A phase space point $\boldsymbol{x} = (\boldsymbol{p}, \boldsymbol{q})$ is mapped onto a phase space point $\boldsymbol{y}$ by

$$\boldsymbol{y} = \boldsymbol{f}_{\mathrm{cl}}(\boldsymbol{x}). \tag{2.1}$$

I have adopted a vector notation in which $\boldsymbol{q} = (q_1, \ldots, q_f)$ denotes the canonical coordinates for f degrees of freedom, and $\boldsymbol{p} = (p_1, \ldots, p_f)$ the conjugate momenta. So far the map can be any function on phase space, but I shall restrict myself to functions that are invertible and differentiable almost everywhere.

Classical maps can arise in many different ways:

- As a Poincaré map of the surface of section from a "normal" Hamiltonian system. Suppose we have a Hamiltonian system with $f = 2$ degrees of freedom ($\boldsymbol{x} = (p_1, p_2, q_1, q_2)$), described by a time-independent Hamiltonian $H(\boldsymbol{p}, \boldsymbol{q})$. Energy is conserved, so the motion in phase space takes place on a $2f - 1 = 3$-dimensional manifold. Many aspects of the motion on the three-dimensional manifold can be understood by looking at an appropriately chosen two-dimensional submanifold. For example, we can look at the plane with one of the canonical coordinates set constant, e.g. $q_2 = q_{20}$. Two coordinates remain free; for example, we may choose p_1, q_1. Such a plane is called surface of section. Whenever the trajectory crosses the plane in the same direction (say with $\dot{q}_2 > 0$), we note the two free coordinates. This yields a series of points $(p_1(1), q_1(1))$, $(p_1(2), q_1(2))$, and so on. We thus have a "sliced" version of the original continuous time trajectory $\boldsymbol{x}(t)$. Whenever $q_2 = q_{20}$, we know in which state the system is. It makes therefore sense to look directly at the map that generates the sequence of points in the plane, the Poincaré map [1].

- The trajectory of a particle that moves in a two dimensional billiard is uniquely defined by the position on the boundary of an initial point and the direction with respect to the normal to the boundary in which the trajectory departs, if we assume that there is no friction and that the particle always scatters off the boundary by specular reflection. All possible trajectories are therefore uniquely encoded in the map that associates with any point on the boundary and any incident angle χ with respect to the normal the following point on the boundary and the corresponding angle. In fact, one can show that the position along the boundary and $\cos \chi$ form a pair of canonically conjugate phase space coordinates and parameterize [16] a surface of section.
- In addition, in the context of periodically driven systems, i.e. systems with a Hamiltonian that is periodic in time, $H(\boldsymbol{x}, t) = H(\boldsymbol{x}, t+T)$, maps arise naturally. Indeed, if we can integrate the equations of motion over one period, we also have the solution for the next period and so on. So it is natural to describe the system stroboscopically by a map that maps all phase space points at time t to new ones at time $t + T$.

Compared with continuous time flows in phase space, maps have several advantages. First of all, one already has an integrated version of the equations of motion. Thus, no differential equations have to be solved to obtain the image of an initial phase space point at a later time. Second, maps can be designed at will and therefore allow one to study "under pure conditions" diverse aspects of chaos. Examples of frequently used maps are the tent map, the baker map, the standard map, Henon's map and the cat map (see [17]). Arnold introduced the sine circle map [18], and May the logistic map [19]. Zaslavsky [20] considered a dissipative generalization of the standard map, and a dissipative version has also been studied for the baker map (see [17]). At present no Hamiltonian system with a standard Hamiltonian $H = T + V$ (where T is the kinetic energy in flat space and V a potential energy) is known that produces hard chaos, i.e. is chaotic everywhere in phase space (see below for a precise definition of chaos), whereas, for example, for the baker map chaos is easily proven.

Maps allow one to study chaos in lower dimensions than do continuous flows. An autonomous Hamiltonian system with one degree of freedom and therefore a two dimensional phase space is always integrable, i.e. it shows regular motion, whereas maps can produce chaos even in a two-dimensional phase space.

All these advantages are particularly favorable if one wants to examine new aspects of chaos such as the connection between classical and quantum chaos in the presence of dissipation, as I shall attempt to do in this book. I shall therefore restrict myself entirely to maps.

2.2 Classical Chaos

Classical chaos is defined as an exponential sensitivity with respect to initial conditions: a system is chaotic if the distance between two phase space points that are initially close together diverges exponentially almost everywhere in phase space. These words can be cast in a more mathematical form by introducing the so-called stability matrix $\mathbf{M}(\boldsymbol{x})$. For a map (2.1) on a $2f$-dimensional phase space, $\mathbf{M}(\boldsymbol{x})$ is a $2f \times 2f$ matrix containing the partial derivatives $\partial f_{\mathrm{cl},i}/\partial x_j$, $i,j = 1,\ldots,2f$, where $f_{\mathrm{cl},i}$ denotes the ith component, or, in shorthand notation, $\mathbf{M}(\boldsymbol{x}) = \partial \boldsymbol{f}_{\mathrm{cl}}/\partial \boldsymbol{x}$. So $\mathbf{M}(\boldsymbol{x})$ is the locally linearized version of $\boldsymbol{f}_{\mathrm{cl}}(\boldsymbol{x})$.

Let $\boldsymbol{x}_0$ be the starting point of an orbit, i.e. a sequence of points $\boldsymbol{x}_0$, $\boldsymbol{x}_1$, ... with $\boldsymbol{x}_{i+1} = \boldsymbol{f}_{\mathrm{cl}}(\boldsymbol{x}_i)$. Then $\mathbf{M}(\boldsymbol{x}_0)$ controls the evolution of an initial infinitesimal displacement $\boldsymbol{y}_0$ from the starting point. After one iteration the displacement is

$$\boldsymbol{y}_1 = \mathbf{M}(\boldsymbol{x}_0)\boldsymbol{y}_0\,, \tag{2.2}$$

and after n iterations we have a displacement

$$\boldsymbol{y}_n = \mathbf{M}(\boldsymbol{x}_{n-1})\mathbf{M}(\boldsymbol{x}_{n-2})\ldots\mathbf{M}(\boldsymbol{x}_0)\boldsymbol{y}_0 \equiv \mathbf{M}^n(\boldsymbol{x}_0)\boldsymbol{y}_0\,. \tag{2.3}$$

The sensitivity with respect to initial conditions is captured by the so called Lyapunov exponent. For an initial direction $\hat{\boldsymbol{u}} = \boldsymbol{y}_0/|\boldsymbol{y}_0|$ of the displacement from the orbit with starting point $\boldsymbol{x}_0$, the Lyapunov exponent $h(\boldsymbol{x}_0,\hat{\boldsymbol{u}})$ is defined as

$$h(\boldsymbol{x}_0,\hat{\boldsymbol{u}}) = \lim_{n\to\infty} \frac{1}{n} \ln|\mathbf{M}^n(\boldsymbol{x}_0)\hat{\boldsymbol{u}}|\,. \tag{2.4}$$

For our map in the $2f$-dimensional phase space there can be up to $2f$ different Lyapunov exponents. However, it can be shown that if an ergodic measure μ_i exists, and this is the case in all examples that will be interesting to us (see next section), the set of Lyapunov exponents is the same for all initial $\boldsymbol{x}_0$ up to a set of measure zero with respect to μ_i [21]. It therefore makes sense to suppress the dependence on the starting point and just call the Lyapunov exponents h_i, $i = 1,2,\ldots,2f$. The fact that they do not depend on $\boldsymbol{x}_0$ is a consequence of the rather general multiplicative ergodic theorem of Furstenberg [22] and Oseledets [23].

Lyapunov exponents are by definition real. The largest one, $h_{\max} = \max(h_1,\ldots,h_{2f})$, is often called "the Lyapunov exponent" of the map. The reason is that for a randomly chosen initial direction $\hat{\boldsymbol{u}}$, the expression in (2.4) converges almost always to $h_{\max}$. In order to unravel the next smallest Lyapunov exponents, special care has to be taken to start with a direction that is in a subspace orthogonal to the eigenvector pertaining to the eigenvalue $h_{\max}$ of the limiting matrix.

We are now in a position to define precisely what we mean by a chaotic map.

Definition: A map is said to be chaotic if the largest Lyapunov exponent is positive, $h_{\max} > 0$.

The sensitivity with respect to initial conditions is hereby defined as a local property in the sense that the two phase space points are initially infinitesimally close together. Of course, the distance between two arbitrary phase space points cannot, typically, grow exponentially forever, since the available phase space volume might be finite. On the other hand, the definition is global in the sense that the total available phase space counts, as the Lyapunov exponents emerge only after (infinitely) many iterations, which for an ergodic system must visit the total available phase space: the Lyapunov exponents are globally averaged growth rates of the distance between two initially nearby phase space points. The definition also applies to maps that have a strange attractor (see next section). In this case the chaotic motion takes place *on* the attractor, and even if the total phase space volume shrinks, two phase space points that are initially close together on the attractor can become separated exponentially fast. If points x in phase space where $\mathbf{M}(x)$ has only eigenvalues with an absolute value equal to or smaller than unity are found on the way, they need not destroy the chaoticity encountered after many iterations.

Lyapunov exponents are related to other measures of classical chaos such as Kolmogorov–Sinai entropy (also called metric entropy) [24, 25] or topological entropy [26]. Since we shall not need these concepts, I refrain from introducing them here and refer the interested reader to the introductory treatment by Ott [17].

Sometimes the above definition is reserved for what is called "hard chaos". A weaker form of chaos arises if some stable islands in phase space exist, i.e. extended regions separated from a "chaotic sea", in which the Lyapunov exponent is not positive. The phase space is then said to be mixed; this situation is by far that most frequently found in nature. It follows immediately that systems with mixed phase space are not ergodic, for if they were, the Lyapunov exponents would be everywhere the same up to regions of measure zero (see the remarks above).

The opposite extreme to chaotic is integrable. Here two initially close phase space points remain close, or at least do not separate exponentially fast. The Lyapunov exponent is zero or even negative. Even though integrable systems such as a single planet coupled gravitationally to the sun (the Kepler problem) and the harmonic oscillator have played a crucial role in the development of the natural sciences, they are very rare. A system can be shown to be integrable iff it has at least as many independent integrals of motion (conserved quantities) as degrees of freedom.

2.3 Ensemble Description

2.3.1 The Frobenius–Perron Propagator

The extreme sensitivity with respect to the initial conditions implies that the description of chaotic systems in terms of individual trajectories is not very useful. Initial conditions can, as a matter of principle, only be known up to a certain precision. If we wanted to measure the position of a particle with infinite precision, we would need some sort of microscope that used light or elementary particles with an infinitely short wavelength and therefore infinite energy. None of this is likely ever to be at our disposal, so it makes sense to accept uncertainties in initial conditions as a matter of principle and try to understand what follows from them.

Uncertainties in the precise state of a system are most easily dealt with in an ensemble description. Instead of one system, we think of very many, eventually infinitely many, copies of the same system. All these copies form an ensemble. The members of the ensemble differ only in the initial conditions, whereas all system parameters (number and nature of particles involved, types and strengths of interaction, etc.) are the same. Instead of talking about the state of the system (that is, the momentary phase space point of an individual member of the ensemble), we shall talk about the *state of the ensemble*. The state of the ensemble is uniquely specified by the probability distribution $\rho_{\rm cl}(\boldsymbol{x},t)$, where t is the discrete time in the case of maps. The probability distribution $\rho_{\rm cl}(\boldsymbol{x},0)$ reflects our uncertainty about the exact initial condition of an individual system, but at the same time it is the precise initial condition of the ensemble. The probability distribution is defined such that $\rho_{\rm cl}(\boldsymbol{x},t)\mathrm{d}\boldsymbol{x}$ is the probability at time t to find a member of the ensemble in the infinitesimal phase space volume element $\mathrm{d}\boldsymbol{x}$ situated at point $\boldsymbol{x}$ in phase space.

In quantum mechanics we are used to thinking that the state of a system is defined by a wave function, which is, however, rather the state of an ensemble. By adopting the ensemble point of view in classical mechanics, the latter looks all of a sudden much more similar to quantum mechanics. In particular, we shall see below that familiar concepts such as Hilbert space and unitary evolution operators exist in classical mechanics as naturally as in quantum mechanics.

In quantum mechanics there is not much alternative to an ensemble description, since to the best of our knowledge there are no hidden variables. Entirely deterministic theories that give the same results as quantum mechanics are possible, but they are non–local. One of them has become known

12   2. Classical Maps

under the name "de Broglie's pilot wave" [27, 28].[1] Classically, the two pictures (individual system vs. ensemble description) are both available and are of course linked to one another. Suppose that the ith individual member of the ensemble has phase space coordinate $\boldsymbol{x}^i(t)$ at time t; then the phase space density, or probability distribution, of the ensemble of M systems is given by

$$\rho_{\text{cl}}(\boldsymbol{x}, t) = \frac{1}{M} \sum_{i=1}^{M} \delta(\boldsymbol{x} - \boldsymbol{x}^i(t)). \tag{2.5}$$

Think of the number M in the limit $M \to \infty$ so that $\rho_{\text{cl}}(\boldsymbol{x}, t)$ can eventually become a smooth function. With the help of (2.5) we can immediately derive the evolution of the phase space density for any map, since $\rho_{\text{cl}}(\boldsymbol{y}, t+1) = (1/M) \sum_{i=1}^{M} \delta(\boldsymbol{y} - \boldsymbol{x}^i(t+1))$. But $\boldsymbol{x}^i(t+1) = \boldsymbol{f}_{\text{cl}}(\boldsymbol{x}^i(t))$, and thus $\rho_{\text{cl}}(\boldsymbol{y}, t+1) = (1/M) \sum_{i=1}^{M} \delta(\boldsymbol{x} - \boldsymbol{f}_{\text{cl}}(\boldsymbol{x}^i(t))) = \int d\boldsymbol{x}\, \delta(\boldsymbol{y} - \boldsymbol{f}_{\text{cl}}(\boldsymbol{x}))(1/M) \sum_i \delta(\boldsymbol{x} - \boldsymbol{x}^i(t))$. So we conclude that

$$\rho_{\text{cl}}(\boldsymbol{y}, t+1) = \int d\boldsymbol{x}\, P_{\text{cl}}(\boldsymbol{y}, \boldsymbol{x})\rho_{\text{cl}}(\boldsymbol{x}, t), \quad P_{\text{cl}}(\boldsymbol{y}, \boldsymbol{x}) = \delta(\boldsymbol{y} - \boldsymbol{f}_{\text{cl}}(\boldsymbol{x})), \tag{2.6}$$

or $\rho_{\text{cl}}(t+1) = P_{\text{cl}}\rho_{\text{cl}}(t)$ for short, if we suppress the phase space arguments. The propagator P_{cl} is most commonly called the "Frobenius–Perron operator" in the context of Hamiltonian systems [32]. For simplicity, I shall use the same name for maps. The connection between $\rho_{\text{cl}}(t+1)$ and $\rho_{\text{cl}}(t)$ can be made explicit by using the property of the Dirac delta function $\delta(\boldsymbol{y} - \boldsymbol{f}_{\text{cl}}(\boldsymbol{x})) = \delta(\boldsymbol{f}_{\text{cl}}^{-1}(\boldsymbol{y}) - \boldsymbol{x})/|\partial \boldsymbol{f}_{\text{cl}}(\boldsymbol{x})/\partial \boldsymbol{x}| = \delta(\boldsymbol{f}_{\text{cl}}^{-1}(\boldsymbol{y}) - \boldsymbol{x})/|\det \mathbf{M}(\boldsymbol{x})|$;

$$\rho_{\text{cl}}(\boldsymbol{y}, t+1) = \frac{\rho_{\text{cl}}(\boldsymbol{f}_{\text{cl}}^{-1}(\boldsymbol{y}), t)}{|\det \mathbf{M}(\boldsymbol{f}_{\text{cl}}^{-1}(\boldsymbol{y}))|}. \tag{2.7}$$

Let me now divide the possible maps into different classes, depending on the allowed values of $|\det \mathbf{M}|$.

2.3.2 Different Types of Classical Maps

A map for which $|\det \mathbf{M}(\boldsymbol{x})| = 1$ for all points $\boldsymbol{x}$ in phase space is locally phase-space-volume-preserving everywhere. In order to have a more handy name I shall call such maps "Hamiltonian", alluding to the fact that all Hamiltonian dynamics is phase-space-volume-preserving. In this book I shall be concerned almost entirely with maps that are *not* Hamiltonian. Obviously,

---

[1] More references and an enlightening discussion can be found in [29]. Note that, strictly speaking, the question whether or not local realistic theories are possible is still not entirely settled. One of the last loopholes (the so-called causality loophole) in the experimental verification of the violation of Bell's inequality that would still allow for a local realistic theory has only recently been closed [30]; another one, the detector loophole, which arises owing to finite detector efficiency, is still considered open and is the subject of strong experimental efforts [31].

this class contains the vast majority of possible maps. Let us therefore divide it further and term maps "dissipative" when the normalized integral of the determinant of the stability matrix over the whole phase space Γ is smaller than unity, $(1/\Omega(\Gamma)) \int_\Gamma |\det \mathbf{M}(\boldsymbol{x})| \, d\boldsymbol{x} < 1$. The volume $\Omega(\Gamma)$ is defined as $\Omega(\Gamma) = \int_\Gamma d\boldsymbol{x}$. The opposite case, $(1/\Omega(\Gamma)) \int_\Gamma |\det \mathbf{M}| \, d\boldsymbol{x} > 1$, defines a "globally expanding map".

The name "dissipative" is motivated by the observation that in a system that dissipates more energy than it receives from outside, the energy shell and therefore the available phase space volume shrink. Of course, the dynamics might still be locally expanding, i.e., for some regions in phase space, the determinant of the stability matrix may be absolutely larger than unity as long as the regions with contracting phase space volume win (see Fig. 2.1). It should be clear that the shrinking of phase space volume does not affect

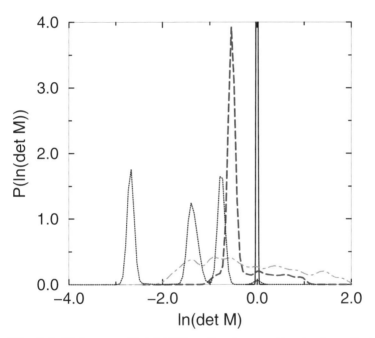

Fig. 2.1. Histogram of $\ln |\det \mathbf{M}|$ on the strange attractor for a dissipative kicked top (see Chap. 6) at $k = 8.0$, $\beta = 2.0$ for increasing dissipation strength. The delta peak at $\ln |\det \mathbf{M}| = 0$ for $\tau = 0$ (*continuous line*) first shifts ($\tau = 0.5$, *dashed line*), then very rapidly broadens ($\tau = 1.0$, *dash–dotted line*), and finally develops a multipeak structure as the attractor covers a smaller and smaller phase space region ($\tau = 2.0$, *dotted line*)

conservation of probability. By the very definition of phase space density, $\int d\boldsymbol{x} \, \rho_{\mathrm{cl}}(\boldsymbol{x}, t) = 1$ always, regardless of the kind of map considered. Indeed, even in the extreme example where all phase space points are mapped to a single point, $\boldsymbol{f}_{\text{cl}}(\boldsymbol{x}) = \boldsymbol{x}_0$ for all $\boldsymbol{x}$, the total probability is conserved, as $\rho_{\text{cl}}(\boldsymbol{x},t) = \delta(\boldsymbol{x}-\boldsymbol{x}_0)$ for all $t \geq 1$ and all initial $\rho_{\text{cl}}(\boldsymbol{x},0)$. So the mapped phase space density is still normalized to one, even though the phase space volume shrinks to zero in one step. This implies, of course, that the mean density in the remaining volume has to increase for dissipative maps. It should therefore be no surprise that dissipative maps lead to invariant phase space structures with dimensions strictly smaller than the phase space dimension, as we shall see in more detail in the next subsection.

The different types of maps allow for different types of fixed points. A fixed point $\boldsymbol{x}_{\text{p}}$ is a point in phase space that is invariant under the map, i.e. $\boldsymbol{f}_{\text{cl}}(\boldsymbol{x}_{\text{p}}) = \boldsymbol{x}_{\text{p}}$. One can also call it a periodic point of period one. A fixed point $\boldsymbol{x}_{\text{p}}$ of $\boldsymbol{f}_{\text{cl}}^2$, i.e. $\boldsymbol{f}_{\text{cl}}(\boldsymbol{f}_{\text{cl}}(\boldsymbol{x}_{\text{p}})) = \boldsymbol{x}_{\text{p}}$, which is not a fixed point of $\boldsymbol{f}_{\text{cl}}$ is called a period-two periodic point, etc. A period-t fixed point has to be iterated t times before it coincides with the starting point. The set of t points that are found on the way (including the starting point) form a periodic orbit. Each of the t points is a period-t periodic point, and one of them is enough to represent the whole periodic orbit. Periodic orbits may be composed of shorter periodic orbits. For example, the iteration of a period-three orbit is a period-six orbit. Periodic orbits that cannot be decomposed into shorter periodic orbits are called primitive periodic orbits or prime cycles [33].

Fixed points play a crucial role in extracting virtually all interesting information from $\boldsymbol{f}_{\text{cl}}$ (as well as from quantum maps, as we shall see). Let us therefore have a closer look at the types of fixed points possible and introduce some terminology that will prove useful later.

Fixed Points of Hamiltonian Maps

The definition of Hamiltonian maps, $|\det \mathbf{M}| = 1$ everywhere, places strong limitations on the nature of the possible fixed points. Note that by the definition of $\mathbf{M}$, $\det \mathbf{M}$ and $\operatorname{tr} \mathbf{M}$ are always real. So the product of the two stability eigenvalues must equal either plus or minus unity. One can easily convince oneself [34] that for $f = 1$ only the following three types are possible:

1. *Hyperbolic fixed points*: both eigenvalues are real and positive; one of them is larger than unity, the other smaller than unity.
2. *Inverse hyperbolic fixed points*: both eigenvalues are real and have absolute values different from unity, but one of them is positive and the other negative.
3. *Elliptic fixed points*: both eigenvalues have absolute values equal to unity and are complex conjugates.

The names originate from the kind of motion that a phase space point in the vicinity of the fixed point will undergo when iterated by the map. In the case of a hyperbolic or an inverse hyperbolic fixed point, the two eigenvectors of $\mathbf{M}$ define a stable and an unstable direction. The former corresponds to the eigenvalue with an absolute value smaller than unity, the latter to the eigenvalue with an absolute value larger than unity. These eigenvectors are tangents to the stable and unstable manifolds. The stable manifold is the set of points that run into the fixed point under repeated forward iteration of the map; the unstable manifold is the corresponding set for backward iteration of the map [34]. A point in the vicinity of the fixed point that is neither on the stable nor on the unstable manifold moves on a hyperbola in the case of a hyperbolic fixed point, but on an ellipse for an elliptic fixed point. Points in the neighborhood of inverse hyperbolic fixed points also move on a hyperbola, but jump from one branch to the other with every iteration of the map.

Fixed Points for Dissipative Maps

Besides the fixed points of Hamiltonian maps, other types can exist here, since $|\det \mathbf{M}|$ can be smaller or larger than unity. Fixed points that have two eigenvalues absolutely smaller than unity are called attracting fixed points (or point attractors); all others are called repelling fixed points (or repellers) [32]. The latter class obviously contains all of the fixed points possible for Hamiltonian maps. Sometimes the term repeller is restricted to fixed points with at least one eigenvalue absolutely larger than unity, and fixed points for which both eigenvalues are absolutely larger than unity are called antiattractors.

2.3.3 Ergodic Measure

Many of the concepts known from quantum mechanics appear in classical mechanics if we talk about classical phase space distributions. Before showing how this comes about, I have to introduce the ergodic measure $\mu_i(\boldsymbol{x})$.

An ergodic measure is an invariant measure that cannot be linearly decomposed into other invariant measures. An invariant measure is a measure that is invariant under the map; $\mu_i(\boldsymbol{f}_{\mathrm{cl}}(\mathcal{M})) = \mu_i(\mathcal{M})$ for any volume $\mathcal{M} \in \Gamma$ in phase space. An invariant measure corresponds to an invariant phase space density $\rho_{\mathrm{cl}}(\boldsymbol{x}, \infty)$ according to $\rho_{\mathrm{cl}}(\boldsymbol{x}, \infty)\mathrm{d}\boldsymbol{x} = \mathrm{d}\mu_i(\boldsymbol{x})$. In general there are many invariant phase space densities. For example, if $\boldsymbol{x}_{\mathrm{p}}$ is a fixed point of the map and if the map is locally phase-space-volume-preserving at $\boldsymbol{x} = \boldsymbol{x}_{\mathrm{p}}$, then $\delta(\boldsymbol{x} - \boldsymbol{x}_{\mathrm{p}})$ is an invariant phase space density (see (2.7)). If there are several fixed points with $|\det \mathbf{M}| = 1$, all linear combinations of delta functions situated on them are invariant phase space densities. If a system is ergodic, however, there is a particular invariant measure that cannot be decomposed into linear combinations of other invariant measures, and furthermore this measure is unique from the definition of an ergodic system. For chaotic Hamiltonian maps this measure is typically a flat measure within a part of phase space selected by the remaining integrals of motion (think, for example, of a Poincaré surface of section that does not show any structure if the map is chaotic). For dissipative chaotic systems one usually encounters a strange attractor, i.e. a self-similar set of phase space points of a dimension strictly smaller than the dimension of the phase space in which it is embedded (see Fig. 6.1). The ergodic phase space density emerges as an invariant state from a generic initial state after infinitely many iterations. That is why I denote it by a time argument of infinity.

The dimension of a strange attractor is a fractal dimension and reflects the self-similarity of the attractor. It is defined as the so-called box-counting dimension, a generalization of the familiar concept of dimension. One studies the scaling of the number $N(\epsilon)$ of little boxes of edge length ϵ needed to cover the attractor with decreasing ϵ. The box-counting dimension is then defined as

$$d = \lim_{\epsilon \to 0} \frac{\ln N(\epsilon)}{-\ln \epsilon}. \tag{2.8}$$

A single point always has $N(\epsilon) = 1$ and therefore vanishing dimension, a line will lead to $N(\epsilon) \propto 1/\epsilon$ and therefore $d = 1$, etc. In Fig. 6.2 I show the dimension of the strange attractor for a dissipative kicked top as a function of dissipation. Overall, the dimension becomes smaller and smaller with increasing dissipation, even though the behavior is not monotonic. Structures very similar to strange attractors also arise from dissipative *quantum* maps, as we shall see in Chap. 7.

2.3.4 Unitarity of Classical Dynamics

Phase space densities obey a (restricted) superposition principle. Suppose $\rho_{\text{cl},1}(\boldsymbol{x})$ and $\rho_{\text{cl},2}(\boldsymbol{x})$ are both valid phase space densities and normalized to unity, then any linear combination $p\rho_{\text{cl},1} + (1-p)\rho_{\text{cl},2}$ with $0 \leq p \leq 1$ is a valid and normalized phase space density as well. The set of all allowed phase space densities does not form a vector space, since the positivity condition $\rho_{\text{cl}}(\boldsymbol{x}) \geq 0$ can prevent the existence of an inverse element for the addition of two densities. Nevertheless, the superposition principle allows us to consider phase space densities as vectors $|\rho_{\text{cl}}\rangle$ in a linear vector space, which then of course contains other elements that do not correspond to physically allowed phase space densities. For example, square-integrable densities $\rho_{\text{cl}}(\boldsymbol{x}, t)$ are elements of the vector space $L^2(\mathbb{R}^{2f})$, and we can expand $\rho_{\text{cl}}(\boldsymbol{x}, t)$ in a complete basis set of (possibly complex) functions. As in quantum mechanics, $\rho_{\text{cl}}(\boldsymbol{x}, t)$ means the vector $|\rho_{\text{cl}}\rangle$ in a certain representation, here the phase space representation, $\rho_{\text{cl}}(\boldsymbol{x}, t) = \langle \boldsymbol{x}|\rho_{\text{cl}}(t)\rangle$. To every ket $|\rho_{\text{cl}}(t)\rangle$ there is a corresponding bra $\langle \rho_{\text{cl}}(t)|$. Since densities are real-valued we have $\langle \rho_{\text{cl}}(t)|\boldsymbol{x}\rangle = \rho_{\text{cl}}(\boldsymbol{x}, t)$. With the help of the ergodic measure, we are in a position to introduce an appropriate scalar product,

$$\langle \rho_{\text{cl},1}|\rho_{\text{cl},2}\rangle = \int d\mu_i(\boldsymbol{x}) \rho_{\text{cl},1}(\boldsymbol{x})^* \rho_{\text{cl},2}(\boldsymbol{x}). \tag{2.9}$$

I have inserted a star for complex conjugation to account for decompositions into complex basis functions.

In complete analogy to the evolution operator of a wave function in quantum mechanics, the Frobenius–Perron propagator of phase space density is a unitary propagator, i.e. $P_{\rm cl}^\dagger P_{\rm cl} = \mathbf{1} = P_{\rm cl} P_{\rm cl}^\dagger$, if the map is Hamiltonian. The Hermitian conjugate of $P_{\rm cl}$ is defined by $\langle \rho_{\rm cl,1} | P_{\rm cl}^\dagger | \rho_{\rm cl,2} \rangle = \langle P_{\rm cl} \rho_{\rm cl,1} | \rho_{\rm cl,2} \rangle$ for all vectors $\langle \rho_{\rm cl,1}|$ and $|\rho_{\rm cl,2}\rangle$. To see that $P_{\rm cl}$ is unitary for Hamiltonian maps, observe that

$$\langle \rho_{\rm cl,1} | P_{\rm cl}^\dagger P_{\rm cl} | \rho_{\rm cl,2} \rangle = \langle P_{\rm cl} \rho_{\rm cl,1} | P_{\rm cl} \rho_{\rm cl,2} \rangle = \int \mathrm{d}\mu_i(\boldsymbol{x}) P_{\rm cl} \rho_{\rm cl,1}(\boldsymbol{x}) P_{\rm cl} \rho_{\rm cl,2}(\boldsymbol{x})$$

$$= \int \rho_{\rm cl}(\boldsymbol{x},\infty) \frac{\rho_{\rm cl,1}(\boldsymbol{f}_{\rm cl}^{-1}(\boldsymbol{x}))}{|\det \mathbf{M}|} \frac{\rho_{\rm cl,2}(\boldsymbol{f}_{\rm cl}^{-1}(\boldsymbol{x}))}{|\det \mathbf{M}|} \mathrm{d}\boldsymbol{x} \,. \quad (2.10)$$

We now switch integration variables to $\boldsymbol{y} = \boldsymbol{f}_{\rm cl}(\boldsymbol{x})$. From the definition of $\rho_{\rm cl}(\boldsymbol{x},\infty)$ as an invariant density we have

$$\rho_{\rm cl}(\boldsymbol{x},\infty) = \frac{\rho_{\rm cl}(\boldsymbol{f}_{\rm cl}^{-1}(\boldsymbol{x}),\infty)}{|\det \mathbf{M}|} \,.$$

Inserting this result, $\mathrm{d}\boldsymbol{y} = \mathrm{d}\boldsymbol{x}/|\det \mathbf{M}|$ and $|\det \mathbf{M}| = 1$ into (2.10), we indeed obtain $\langle \rho_{\rm cl,1} | P_{\rm cl}^\dagger P_{\rm cl} | \rho_{\rm cl,2} \rangle = \langle \rho_{\rm cl,1} | \rho_{\rm cl,2} \rangle$. Note that, as a decisive ingredient, we have used $|\det \mathbf{M}| = 1$ everywhere.

A consequence of the unitarity of $P_{\rm cl}$ for Hamiltonian maps is that the quantity $\int \mathrm{d}\mu_i(\boldsymbol{x}) \rho_{\rm cl}^2(\boldsymbol{x},t)$ is conserved, as is in quantum mechanics the norm of a state vector, $\int \mathrm{d}\boldsymbol{x} |\psi(\boldsymbol{x})|^2$. Nevertheless, the issue of unitarity has traditionally played a far less important role in classical dynamics than in quantum mechanics. There are of course historical reasons for this, as classical mechanics was first formulated as a dynamics of mass points and not of ensembles, and most effort has been concentrated on understanding the former. But a more intrinsic reason is certainly that an important part of classical dynamics does *not* take place in Hilbert space. For chaotic systems finer and finer structures appear with evolving time, leading to generalized eigenstates of the Frobenius–Perron operator in the form of distributions or worse, as we shall see in the next subsection.

2.3.5 Spectral Properties of the Frobenius–Perron Operator

As long as the Frobenius–Perron operator is a unitary operator acting on a Hilbert space only, it has a spectrum entirely on the unit circle. The spectrum may contain a discrete part, namely eigenvalues $\lambda_n = \exp(\mathrm{i}\omega_n t)$ with real ω_ns, and a continuous part. Iff $\lambda = 1$ is simply degenerate, the system is ergodic. And iff $\lambda = 1$ is simply degenerate and the only (discrete) eigenvalue, the system is a mixing system. Mixing systems have, however, necessarily a continuous spectrum as well [32]. The corresponding eigenstates are generalized eigenstates. They are not part of the Hilbert space, and they are not even functions but linear functionals.

This may sound like a contradiction to the reader unfamiliar with the mathematical subtleties of spectral theory, but the point is that the spectrum is defined via the resolvent $R(z) = 1/(z - P_{\mathrm{cl}})$, where z is a complex number. The resolvent $R(z)$ is said to exist for a given z if $R(z)|\psi\rangle$ is defined for all vectors $|\psi\rangle$ in Hilbert space, and if for every vector $|\psi\rangle$ in Hilbert space $R(z)|\psi\rangle$ is again in Hilbert space. A point z_n is defined as an element of the point spectrum (i.e. is a discrete eigenvalue of P_{cl}) if $R(z)$ does *not* exist for $z = z_n$. A point z is an element of the continuous spectrum if $R(z)$ exists but is not bounded. That means we can find a series of vectors in Hilbert space such that, for all elements of the series, $R(z)|\psi\rangle$ is defined and again a vector in Hilbert space, but the norm of $R(z)|\psi\rangle$ diverges within the series. A good example is the familiar position operator $\hat{x}$ in quantum mechanics, which has a purely continuous spectrum. Its generalized eigenstates are delta functions $\delta(x - x_0)$ centered at arbitrary positions x_0. The delta function is not part of Hilbert space, since it is not square integrable, but it may be approximated better and better by a series of narrowing Gaussian peaks that are in Hilbert space. Therefore the corresponding resolvent exists but is not bounded.

The spectrum on the unit circle is also called the spectrum of real frequencies of P_{cl} (since ω is real). Unfortunately, this spectrum does not say much about the transient behavior of a system on the way to its long-time limit. The transient behavior shows up, for example, in correlation functions of observables and typically contains exponential decays, exponentially damped oscillations or exponentials combined with powers of time. Such behavior can be understood from the spectral properties of P_{cl} if we extend the spectral analysis to complex frequencies ω. It is clear that if P_{cl}^t has an eigenvalue $\exp(\mathrm{i}\omega t)$ with complex ω we can expect an exponentially damped oscillation, where the imaginary part of ω sets the damping timescale and the real part the timescale for the oscillation. The eigenvalues of P_{cl} with complex ω are commonly called Ruelle resonances [21, 35, 36, 37, 38] or Policott–Ruelle resonances [39, 40]. The corresponding eigenstates are again generalized eigenstates, i.e. they do not live in Hilbert space. The Ruelle resonances for a map that has a single fixed point are directly connected with the stabilities of the fixed point [32]. In general one can calculate at least the leading Ruelle resonances via trace formulae and an appropriate zeta function, as I shall show in more detail in Sect. 7.2.3. The Ruelle resonances play an important role for the spectrum of the *quantum* propagator for a corresponding dissipative quantum map, as we shall see.

2.4 Summary

In this chapter I have introduced some basic concepts of classical chaos, focusing on dissipative maps of phase space on itself. Chaos has been defined via Lyapunov exponents, and we have seen a Hilbert space structure arise in classical mechanics, just by going over to an ensemble description. The

Frobenius–Perron propagator of phase space density was introduced, and its invariant state and some of its spectral properties were discussed. I shall later come back to these properties in the case of dissipative quantum maps.

3. Unitary Quantum Maps

After the crash course on classical maps and classical chaos in the preceding chapter, let us now have a look at the corresponding concepts in quantum mechanics. I shall introduce in this chapter about unitary quantum maps the object of choice for studying chaos in ordinary, i.e. nondissipative quantum mechanics. A standard example, namely a kicked top, will serve as a useful model, not only in this chapter, but for the rest of this book. We shall see how a classical map emerges from the quantum map, and identify signatures of chaos in the quantum world. With the Van Vleck propagator and Gutzwiller's trace formula, we shall also encounter for the first time semiclassical theories that try to bridge the gap between chaos in the classical realm and in the quantum world. The generalization of these semiclassical theories to dissipative dynamics will be the main topic of later chapters.

3.1 What is a Unitary Quantum Map?

A quantum map is a map that maps a quantum mechanical object such as a wave function or a density matrix. In this chapter we shall consider unitary quantum maps. Unitary quantum maps map a state vector by a fixed unitary transformation F in Hilbert space,

$$|\psi(t+1)\rangle = F|\psi(t)\rangle, \quad FF^\dagger = \mathbf{1} = F^\dagger F. \tag{3.1}$$

By the argument t I denote a discrete time as in the case of classical maps, $t = 0, 1, 2, \ldots$

Similarly to classical maps, quantum maps arise in a variety of contexts. For example, if a system has a Hamiltonian that is periodic in time with period T, $H(t) = H(t+T)$ (for the moment let t denote a continuous time), the evolution operator of the state vector over one period is given by

$$U(T) = \left[\exp\left(-\frac{\mathrm{i}}{\hbar}\int_0^T H(t)\,\mathrm{d}t\right)\right]_+, \tag{3.2}$$

where the subscript "+" denotes positive time ordering, i.e.

$$[A(t)B(t')]_+ = \begin{cases} A(t)B(t') & \text{for } t > t' \\ B(t')A(t) & \text{for } t < t' \end{cases}. \tag{3.3}$$

22   3. Unitary Quantum Maps

If we are interested only in a stroboscopic description of the quantum dynamics, i.e. in the state vectors at discrete times nT, we can use (3.1) with $F = U(T)$. Owing to the periodicity of $H(t)$, the evolution operator for one period is always the same, so that $U(nT) = U(T)^n = F^n$. The matrix F is called the Floquet matrix [41]. It contains all the information about the stroboscopic dynamics. Typical situations where the Hamiltonian is a periodic function of time are the interaction of a laser with atoms, electron spin resonance, nuclear magnetic resonance (see P. Hänggi in [42]), or driven chemical reactions [43].

Many of the classical maps that have played an important role in understanding classical chaos have been quantized. In particular, there is a quantum baker map [44, 45, 46, 47] on the torus as well as on the sphere [48], and a quantum version of the standard map, the kicked rotator [49, 50, 51, 52, 53, 54, 55, 56]. Another example where quantum maps arise is that of quantum Poincaré maps [57, 58].

As a unitary matrix, F has unimodular eigenvalues $\exp(i\varphi_j)$, $j = 1, \ldots, N$, where N is the dimension of the Hilbert space in which F acts. The "eigenphases" are called quasi-energies. In a system with a constant Hamiltonian H they are related by $\varphi_j = -E_j t/\hbar$ (modulo 2π) to the true eigenenergies E_j of H. We shall be particularly interested in quantum maps that have a classical limit. So it should be possible to derive a well-defined and unique classical map in phase space from the quantum map in the limit where an "effective $\hbar$" in the system approaches zero. The effective $\hbar$ can be, for example, the inverse dimension of the Hilbert space. The example presented in the next section will clarify this point. In Sect. 3.3 I shall discuss briefly how chaos manifests itself on the quantum mechanical level for unitary quantum maps, and the rest of this chapter will be devoted to semiclassical methods that bridge the gap between classical and quantum chaos.

In Chap. 6 we shall consider dissipative quantum maps, for which the description by a state vector is not sufficient and one has to go over to density matrices.

3.2 A Kicked Top

Let me introduce in this section the example of a kicked top, a simple but very fruitful example of a unitary quantum map. I shall use this map throughout this book to illustrate various aspects of quantum chaos. Even for dissipative quantum maps, the kicked top will play an important role.

The dynamical variables of a top [59, 60] are the three components J_x, J_y, and J_z of an angular momentum $\boldsymbol{J}$. The origin of the name "kicked top" can be clearly seen from the Hamiltonian,

$$H(t) = \hbar \left(\frac{k}{2JT} J_z^2 + \beta J_y \sum_{n=-\infty}^{\infty} \delta(t - nT) \right). \tag{3.4}$$

The first term, which is independent of time, tries to align the angular momentum in the plane $J_z = 0$. In solid-state physics such a term can arise from nonisotropic crystal fields, for example in crystals that have an easy plane of magnetization [61]. The second term describes a periodic kicking of the angular momentum. The time evolution operator $F = U(T)$ that maps the state vector from its value at time $t = 0_-$ to time T_- follows from (3.2);

$$F = e^{-i(k/2J)J_z^2} e^{-i\beta J_y} . \tag{3.5}$$

The dynamics generated by F conserves the absolute value of $\boldsymbol{J}$, i.e. $\boldsymbol{J}^2 = j(j+1) = \text{const}$, where j is a positive integer or half-integer quantum number. The classical limit is formally attained by letting the quantum number j approach infinity. Indeed, one can measure the degree to which an angular momentum is a classical quantity by counting the number of angular-momentum quanta $\hbar$ that it contains. The quantum number j is just this number of quanta. It will turn out that js of the order of 5–10 can already lead to rather classical behavior; and 5–10 is still far away from the angular momenta which we encounter in the classical mechanics of everyday life, which have values of j of the order of 10^{34}.

The surface of the unit sphere $\lim_{j\to\infty}(\boldsymbol{J}/j)^2 = 1$ becomes the phase space in the classical limit. It is two-dimensional, or, in other words, we have but a single degree of freedom. Besides j, it is convenient to introduce also $J = j + 1/2$ since this parameter simplifies many formulae.

A convenient pair of phase space coordinates is

$$\mu \equiv J_z/J = \cos\theta = p, \quad \phi = q, \tag{3.6}$$

where the polar and azimuthal angles θ and ϕ define the orientation of the classical angular-momentum vector with respect to the J_x and J_z axes. To see that $\cos\theta$ and ϕ are canonically conjugate, we must verify that the Poisson bracket

$$\{\cos\theta, \phi\} = 1 \tag{3.7}$$

leads to the correct quantum mechanical commutator if we replace the classical variables $x \equiv J_x/J = \sin\theta\cos\phi$, $y \equiv J_y/J = \sin\theta\sin\phi$ and $z \equiv J_z/J = \cos\theta$ by the corresponding quantum mechanical operators $\hat{J}_x/J$, $\hat{J}_y/J$ and $\hat{J}_z/J$, and the Poisson bracket by a commutator [60]. This is indeed the case, since from (3.7) we can deduce for any $f(z), g(\phi)$ the Poisson bracket $\{f(z), g(\phi)\} = f'(z)g'(\phi)$ and thus, from the definition of x, y and z in terms of the angles θ and ϕ, that $\{x, y\} = -z$. We recover the familiar angular-momentum commutation relations $[\hat{J}_x, \hat{J}_y] = i\hbar\hat{J}_z$ if we replace the Poisson bracket with the commutator according to $\{,\} \to (i/J)[,]$. The latter relation shows that $\hbar$ scales as $1/J$. In the following we shall set $\hbar = 1$ and keep $1/J$ as a measure of the classicality, the classical limit corresponding to $1/J \to 0$. The hats on the operator symbols will be dropped, as long as the distinction is clear from the context.

Owing to the conservation of $\boldsymbol{J}^2$ the Hilbert space decomposes into $(2j+1)$-dimensional subspaces defined by $\boldsymbol{J}^2 = j(j+1)$. The quantum dynamics is confined to one of these subspaces according to the initial conditions. Since the classical phase space contains $(2j+1)$ states, we see once more that Planck's constant $\hbar$ may be thought of as represented by $1/J$.

The angular-momentum components are generators of rotations. The unitary evolution generated by the Floquet matrix (3.5) first rotates the angular momentum by an angle β about the y axis and then subjects it to a torsion about the z axis. The latter may be considered as a nonlinear rotation where the rotation angle is itself proportional to J_z. The dynamics is known to become strongly chaotic for sufficiently large values of k and β, whereas either $k = 0$ or $\beta = 0$ lead to integrable motion [60]. For a physical realization of this dynamics one may think of $\boldsymbol{J}$ as a Bloch vector describing the collective excitations of a number of two-level atoms, as one is used to in quantum optics. The rotation can be brought about by a laser pulse of suitably chosen length and intensity, and the torsion by a cavity that is strongly off resonance from the atomic transition frequency [62].

3.3 Quantum Chaos for Unitary Maps

The classical definition of chaos cannot be applied to quantum mechanics: the stability matrix **M** is not defined, since quantum mechanics does not know the notion of a trajectory in phase space. The quantum mechanical trajectories that one might think of are the trajectories of the state vector in Hilbert space. But, of course, there can be no such thing as exponential sensitivity with respect to the initial conditions, since the unitary time evolution preserves the angles and distances between different states. It is therefore not possible to distinguish between chaotic and integrable dynamics from the sensitivity with respect to the initial conditions of the motion in Hilbert space. But neither is this is possbible in classical mechanics if the latter is formulated in terms of a dynamics of state vectors in a Hilbert space describing probability distributions. In classical mechanics we have the additional notion of the trajectories of single particles in phase space, which is absent in quantum mechanics. So we should look for another quantum mechanical criterion for chaoticity.

Many quantum mechanical criteria have been proposed. To do justice to all of them is impossible in the present short section. Furthermore, I am mostly interested in quantum chaos in the presence of dissipation, which will be discussed in more detail in Chaps. 6 and 7. I shall therefore just state some of the criteria, comforting the disappointed reader with [60].

A simple criterion that is closely related to the definition of classical chaos is the sensitivity of quantum trajectories with respect to changes in control parameters [63, 64]. The overlap between two state vectors that are propagated by maps with slightly different control parameters decreases exponentially with time in the chaotic case, but not in the regular case. This observation was also used later in terms of phase space densities for classical chaos [65], and embedded in a broader information-theoretical framework. Instead of a perturbation of the system through a small change of a system parameter, the distribution of Hilbert space vectors that arises from interaction with an environment was studied [66].

The way a system interacts with its environment has also been used by Miller and Sarkar [67] to distinguish chaotic quantum systems from integrable ones. These authors have shown that the quantum mechanical entanglement between two coupled kicked tops increases linearly in time with a rate proportional to the sum of the positive Lyapunov exponents.

Miller and Sarkar also generalized the classical concept of entropy production as a criterion for chaoticity to the quantum world [68]: the von Neumann entropy tr $\rho \ln \rho$ increases much more rapidly for chaotic systems than for integrable ones.

Definitely the most popular criterion for chaoticity in the quantum world is the based on the so-called random-matrix conjecture. Put forward by Bohigas et al. [5, 69] et al. and Berry [6], the conjecture states that the energy spectra and eigenstates of quantum mechanical systems with a chaotic classical limit have special statistical properties that distinguish them from systems with an integrable classical limit. The statistical properties correspond to those of certain random matrices, where the randomness is restricted only by general symmetry requirements. Owing to this conjecture, there exists a close link between the quantum theory of classically chaotic systems and random-matrix theory (RMT). The latter was invented in the 1950s by Wigner, Dyson and others in order to describe the overwhelmingly complex spectra of heavy nuclei [70, 71]. Even though a rigorous proof has still not been published, impressive numerical and experimental evidence for the correctness of the conjecture has been collected.

Dyson's ensembles of unitary matrices, the so-called circular ensembles, are relevant to maps. As for Hermitian matrices, one distinguishes three classes by means of the same symmetry considerations: the circular orthogonal ensemble (COE), the circular unitary ensemble (CUE) and the circular symplectic ensemble (CSE). These names originate from the fact that the probability to find a certain unitary matrix in the ensemble is invariant under orthogonal, unitary and symplectic transformations, respectively. The COE applies to systems with an antiunitary symmetry T that squares to unity, $T^2 = 1$, which means that the Floquet matrix of the physical system has to be covariant under T, i.e. $TFT^{-1} = F^\dagger$ [60]. A typical antiunitary symmetry is conventional time reversal symmetry. For systems without an antiunitary symmetry the CUE is the relevant ensemble, and for systems with an antiunitary symmetry that squares to -1, the CSE is relevant.

A detailed account of the properties of these ensembles would be beyond the scope of the present book, and detailed overviews exist elsewhere [60,

72, 73, 74, 75]. Here I would just like to point out a key consequence that is common to all the ensembles: level repulsion. The probability to find two levels (or eigenphases, as for the case of the circular ensembles) close together is strongly suppressed compared with uncorrelated random sequences. This fact is expressed in the N-point joint probability distribution function of eigenphases $P(\varphi_1, \ldots, \varphi_N)$, which is given by

$$P(\varphi_1, \ldots, \varphi_N) = C_{N\beta} \prod_{k \neq l} |e^{i\varphi_k} - e^{i\varphi_l}|^\beta, \tag{3.8}$$

where $\beta = 1, 2$ and 4 for the COE, CUE and CSE, respectively, N is the dimension of the random matrices and $C_{N\beta}$ is a normalization constant. Other arbitrary statistics can be derived from the joint probability distribution. The most widely used for practical purposes is the nearest-neighbor-spacing distribution $P(s)$, which is the distribution of distances $\varphi_{i+1} - \varphi_i$ between neighboring eigenphases φ_i. With the first and zeroth moments normalized to one, it has the same form for both the circular and the Gaussian ensemble for $N \to \infty$ [73]. The function is very well approximated by a simple formula obtained from the $N = 2$ case, the so-called Wigner surmise,

$$P(s) = A_\beta s^\beta e^{-B_\beta s^2}, \tag{3.9}$$

where the constants A_β and B_β ensure the right normalization. Thus, the probability to find two eigenphases close together vanishes according to a power law at small distances; the power is given by the symmetry of the ensemble. Integrable classical dynamics leads generically to uncorrelated spectra [76, 77]. Recently it has become clear that the three different symmetry classes introduced by Wigner and Dyson are not the only possible ones [78]. And even within the same symmetry class, different stable statistics are possible (i.e. statistics that are independent of the system size for actual physical systems). For example, in disordered systems it is well known that at the Anderson metal–insulator transition in three dimensions another universal type of statistics exists [79, 80], and there are in fact infinitely many such types, depending on how the boundary conditions and the aspect ratio of the sample are chosen [81]. In Chap. 7 we shall encounter another ensemble, which is meant to describe nonunitary propagators.

Let me finish this section by pointing out that the ideas of quantum chaos have been fruitful for classical systems, too. In particular, the statements about level statistics and energy eigenstate statistics seem to apply also to classical wave problems, for example for microwave billiards [82] and acoustic waves [83], even though the issue of universal parametric correlations in experimentally obtained spectra is not settled yet. Further applications exist for driven stochastic systems, where a Fokker–Planck equation with a periodic coefficient arises [84].

3.4 Semiclassical Treatment of Quantum Maps

There is no sharp boundary between classical and quantum mechanics. Quantum mechanical effects become more and more visible if typical actions in a system become comparable to $\hbar$ and if the coherence of wave functions is not disturbed. In our example of a kicked top we can make a continuous transition between quantum mechanics and classical mechanics by increasing the value of the angular-momentum quantum number j. Nevertheless, we have seen that the signatures of chaos are very different in quantum mechanics and in classical mechanics. Semiclassical theories try to bridge the gap between the two extremes.

In 1928 Van Vleck introduced a semiclassical propagator, which, apart from besides brute numerical solution of the Schrödinger equation, has been until today the only way to tackle the quantum mechanics of systems that are classically chaotic. The semiclassical methods that we are interested in here use classical input to calculate approximatively the quantum mechanical quantities such as propagators, transition matrix elements, the density of states and its correlation function. I shall not derive these methods here, since many good descriptions exist elsewhere [74, 85, 86, 87, 88]. The following two subsections are intended rather as an overview of some important concepts that I shall generalize later on for dissipative systems.

3.4.1 The Van Vleck Propagator

The propagator invented by Van Vleck approximates the time evolution operator (3.2) for the Schrödinger equation [89]. Written in a given representation, say the position basis, the time evolution operator is a matrix, $\langle x|U(T)|x'\rangle$. The labels x' and x define the starting and end points of one or several classical trajectories σ that run from x' to x within the time T, and one sums over all such trajectories. Each contributes a complex number, whose phase is basically the classical action $S(x, x'; T)$ in units of $\hbar$ accumulated along the trajectory. The amplitude is related to the stability of the trajectory;

$$\langle x|U(T)|x'\rangle \simeq \frac{1}{\sqrt{2\pi\hbar}} \sum_\sigma \sqrt{|\partial_x \partial_{x'} S_\sigma(x, x'; T)|} e^{i\left(S_\sigma(x,x';T)/\hbar - (\nu_\sigma + 1/2)(\pi/2)\right)}. \quad (3.10)$$

The integer ν_σ, the so-called Morse index counts the number of caustics encountered along the trajectory σ [88, 90, 87].

For maps, a corresponding general propagator with the same structure was derived by Tabor [91]. I shall give it right away for the kicked top (3.5); we shall need this result in the following [92]. It is most easily written in the momentum basis, as we have identified $\mu = m/J$ as the classical momentum, where m is the J_z quantum number ($J_z|m\rangle = m|m\rangle$). The torsion part is already diagonal in this representation and just leads to a phase factor. The rotation about the y axis gives rise to Wigner's d function [93, 94]. Accounting for the fact that $1/J$ plays the role of $\hbar$, one finds

$$\langle n|F|m\rangle = \frac{(-1)^j}{\sqrt{2\pi J}} \sum_\sigma \sqrt{|\partial_\nu \partial_\mu S_\sigma(\nu,\mu)|} e^{i(JS_\sigma(\nu,\mu)-\nu_\sigma \pi/4)}, \qquad (3.11)$$

where $\nu = n/J$. Explicit expressions for the action $S(\nu,\mu)$ can be found in [92], where a geometrical interpretation of the classical dynamics was also given. We shall, however, never need the explicit form of S. Rather, the important features are its generating properties, which connect the initial and final canonical coordinates ϕ^i and ϕ^f of the trajectory to partial derivatives of the action,

$$\partial_\nu S_\sigma(\nu,\mu) = -\phi^f_\sigma(\nu,\mu),$$
$$\partial_\mu S_\sigma(\nu,\mu) = \phi^i_\sigma(\nu,\mu), \qquad (3.12)$$

for each trajectory σ.

3.4.2 Gutzwiller's Trace Formula

Gutzwiller used the Van Vleck propagator to express the trace of the Green's function of a quantum system in terms of purely classical quantities [3, 4]. The interest in the trace of the Green's function lies in the fact that its imaginary part gives the spectral density. A corresponding formula for maps was derived by Tabor [91]. The situation here is in principle somewhat simpler, since the traces of powers of F can be used directly to calculate its eigenvalues, as we shall see. The decisive ingredients in these "trace formulae" are periodic points. For each periodic point (p.p.) of $\boldsymbol{f}^t_{\text{cl}}$ (where $t \in \mathbb{N}$ is again the discrete time), one obtains a complex number in which the action S accumulated in the periodic orbit of length t starting at the periodic point determines the phase. The amplitude depends on the trace of the total stability matrix $\mathbf{M}$ of the periodic point, defined in Chap. 2. Again, I shall not derive the trace formula here, but just cite the result [91];

$$\operatorname{tr} F^t = \sum_{\text{p.p.}} \frac{e^{i(JS-\mu\pi/2)}}{|2-\operatorname{tr}\mathbf{M}|^{1/2}}. \qquad (3.13)$$

All ingredients in this formula are canonical invariants. The integer μ (commonly called the Maslov index) differs typically from the Morse index in the propagator. It has the simple topological interpretation of a winding number [95, 96].

Traces of propagators are not very interesting per se. From a physical point of view, we are much more interested in the spectrum of the propagator, or at least in statistical properties like spectral correlations. There are several ways in which one can learn something about the spectrum of unitary quantum maps from the traces $\operatorname{tr} F^t$. The first way is a connection between the spectral form factor and the absolute squares of the traces. The spectral form factor is the Fourier transform of the spectral-density correlation function, and has played an important role in semiclassical attempts to prove the RMT conjecture. Let me focus here on another connection, however, since it gives – at least in principle – direct access to the spectrum even for dissipative quantum maps.

Suppose we know F in an N-dimensional matrix representation. The nth trace, $\operatorname{tr} F^n$, is given in terms of the N eigenvalues λ_i by $t_n \equiv \operatorname{tr} F^n = \sum_{i=1}^{N} \lambda_i^n$. If we know the traces for $n = 1 \ldots N$, we have N nonlinear equations for the N unknown eigenvalues. The problem of inverting these equations, i.e. expressing the eigenvalues as functions of the traces, was solved long ago by Sir Isaac Newton. He related the coefficients a_n of the characteristic polynomial

$$\det(F - \lambda) = \sum_{n=0}^{N} (-\lambda)^n a_{N-n} \tag{3.14}$$

to the traces of F. The derivation is based on expressing the characteristic polynomial as

$$\det(F - \lambda) = \exp(\operatorname{tr} \ln(F - \lambda))$$

and subsequent expansion in a power series. The details of the derivation can be found in [88]. Let me restrict myself to quoting the result: the coefficient a_0 equals unity, and $a_N = \det F$. The other coefficients are calculated most efficiently by the recursion formula

$$a_n = \left((-1)^{n-1} t_n + \sum_{m=1}^{n-1} (-1)^{m-1} t_m a_{n-m} \right) / n. \tag{3.15}$$

After construction of the polynomial, we can solve it numerically for its roots and obtain the eigenvalues of F. In the case of infinite-dimensional operators the above recursion formulae are known as the Plemelj–Smithies recursion (see appendix F in [14]).

3.5 Summary

With unitary quantum maps, I have introduced in this chapter a quantum mechanical analogue of the classical maps of Chap. 2. More precisely, unitary quantum maps correspond to what was called Hamiltonian classical maps in Chap. 2, i.e. maps that are phase-space-volume-conserving everywhere. I have mentioned a few manifestations of chaos in unitary quantum maps. And we have seen, with the Van Vleck propagator and Gutzwiller's trace formula, how classical information can be used to gain insight into the quantum mechanical behavior. These concepts will be generalized in the following chapters to situations where dissipation cannot be neglected.

4. Dissipation in Quantum Mechanics

Let me review in this chapter how dissipation can be dealt with in quantum mechanics. After general preparatory remarks in the first section, I shall focus on a particular example and show how a dissipative propagator can be approximated in a systematic way semiclassically. I shall use the same dissipation mechanism for dissipative quantum maps later on as a relaxation process, so that the propagator derived in this chapter will find an important application.

4.1 Generalities

Dissipative systems can lose energy owing to a coupling to an external world. Schrödinger's equation, on the other hand, was invented for Hamiltonian systems, i.e. systems that conserve total energy, or at least have a well-defined time-dependent Hamiltonian. Besides dissipation of energy, which arises even in classical mechanics, the coupling to the external world also leads in general to the purely quantum mechanical effect of decoherence, which will be described in more detail in the next chapter.

There have been many different approaches to dissipation in quantum mechanics [97]. The one which is most appealing from a physical point of view is the so-called Hamiltonian embedding. Here, the dissipative system is understood as part of a larger Hamiltonian system which has, in general, very many degrees of freedom. Dissipation arises because the system of interest can exchange energy with the rest of the larger system, usually called the "heat bath" or the "environment". The total system is assumed to be closed, so that the total energy is conserved. It can therefore be adequately described by ordinary quantum mechanics, i.e. a Schrödinger equation for a many-particle wave function. The total Hamiltonian is composed of a part which describes the system of interest without dissipation, H_s, the Hamiltonian for the heat bath, H_b, and a coupling term H_{int}, which couples the system to the environment. This approach is appealing from a physical point of view, since no elementary particle, atom, molecule or other system exists alone and by itself in nature. It always couples to the rest of the universe, for example to electromagnetic waves by scattering of photons, to air molecules, or even to faraway galaxies via the omnipresent gravitational forces.

4. Dissipation in Quantum Mechanics

Depending on the system under consideration, one or other group of environmental degrees of freedom may be more important. For example, an ion embedded in a solid-state crystal will feel most dominantly its neighboring ions and electrons. The most relevant degrees of freedom to which it can dissipate energy are therefore lattice oscillations of the crystal or excitations of the electrons via electromagnetic interactions. It is one of the strengths of the Hamiltonian embedding approach that a more or less realistic model can be made not only of the system of interest but also of its environment and the coupling to the environment. Dissipation can manifest itself very differently depending on the nature of the heat bath; and the Hamiltonian embedding offers a microscopic picture of the possible effects.

In the presence of dissipation, it is natural to describe the system not by a wave function but by a density matrix. This allows for more general initial conditions, for instance an initial condition where the heat bath is in thermal equilibrium at temperature T and is therefore described by an initial density matrix $W_b(0) = e^{-\beta H_b}/Z$, where $\beta = 1/k_B T$, $Z = \text{tr}\, e^{-\beta H_b}$ and k_B is the Boltzmann constant. The time evolution of the total density matrix $W(t)$ is given by the von Neumann equation,

$$i\hbar \frac{d}{dt} W(t) = [H, W(t)] . \tag{4.1}$$

Suppose that at a time t we want to measure the observable A of the system. So A is an operator that acts only on the system Hilbert space $\mathcal{H}_s$. The key observation is that when we measure A, the degrees of freedom of the environment remain unobserved, i.e. our measurement leaves the heat bath part of the total wave function as it is. In other words, in the total system we measure $A \otimes \mathbf{1}_b$, where $\mathbf{1}_b$ denotes the unit operator in the environmental part of the Hilbert space. According to the postulates of quantum mechanics, the expectation value of A at time t is given by

$$\langle A(t) \rangle = \text{tr}_{\text{tot}}[A \otimes \mathbf{1} W(t)] = \text{tr}_s[A\rho(t)], \tag{4.2}$$

with the so-called reduced density matrix defined by

$$\rho(t) = \text{tr}_b W(t). \tag{4.3}$$

It is a "reduced" density matrix, because the environmental degrees of freedom have been traced out, as denoted by tr_b. The time development of $\rho(t)$ gives an effective picture of the dissipative dynamics of the system of interest in contact with the chosen heat bath. The main task is therefore to find and solve the effective equation of motion for $\rho(t)$ from the definition (4.3) and the evolution of $W(t)$ according to (4.1). This has been achieved explicitly only for a very few models [97]. For a long time interest was focused on models with infinitely many but exactly solvable degrees of freedom for the environment, in particular, heat baths consisting of infinitely many harmonic oscillators [98, 99, 100, 101]. An exact solution has been obtained for the dissipative harmonic oscillator [102], and very good understanding has been achieved for the dissipative two-state system (for a review see [103] or [97]) as well as for various other models in solid-state physics, such as damped hopping of a particle in a one-dimensional crystal lattice [104] and even rotational tunneling of small molecular groups in a molecular crystal [105]. Recently, models in which the heat bath is formed by very few (or a even only one), but chaotic degrees of freedom have also attracted attention [106].

In general, the equation of motion for $\rho(t)$ is a complicated integro-differential equation. Physically, this structure arises owing to memory effects. As the heat bath "remembers" the trajectory of a particle for a certain time, the behavior of $\rho(t)$ depends on its earlier history. However, if we are interested only in timescales that are much longer than the memory time of the heat bath, the equation of motion for $\rho(t)$ can be greatly simplified. We are then led to a so-called Markovian master equation, in which the time derivative of $\rho(t)$ depends only on $\rho(t)$ at the same time t, and not on its values at earlier times. Instead of a complicated integro-differential equation, we get a simpler differential equation.

In this book I am not primarily interested in the modeling of a particular form of dissipation as realistically as possible. Rather, I focus on the combined effects of dissipation and chaos, and on semiclassical approaches to quantum chaos in the presence of dissipation. I shall therefore restrict myself to dissipative processes that can be described by simple Markovian master equations, as a good compromise between technical feasibility and physical reality. Markovian master equations have a broad range of application and well-defined limits of applicability, and are very frequently encountered in quantum optics.

Let us have a look at a particular example that will serve for the rest of this book as model damping mechanism.

4.2 Superradiance Damping in Quantum Optics

4.2.1 The Physics of Superradiance

Consider a cloud of N two-level atoms, all initially excited into the upper state. Sooner or later each atom will emit a photon by spontaneous emission. Let $\tau_{\rm sp}$ be the characteristic time for this to happen. As long as there is no coupling between the atoms, each atom will emit its photon independently of the others and in an arbitrary direction, and so the intensity of the emitted light decays exponentially with time, $I(t) = I_0 \exp(-t/\tau_{\rm sp})$. This is the essence of ordinary fluorescence.

Now suppose that the atoms are inside an optical cavity which supports modes of the electromagnetic field at discrete frequencies ω_i. Let all atoms be in resonance with a single mode at the frequency ω_0, i.e. the energy $\hbar\omega_0$ of a photon in the resonant electromagnetic mode matches the energy separation between the two levels in the atoms. The atoms are thus coupled via the cavity mode. What now happens is the following. A first atom emits its photon by spontaneous emission. However, this photon does not escape right away, but is fed into the resonant cavity mode, where it can immediately interact with all the other atoms. It induces emission in a second atom, whose photon goes again into the cavity mode. The two photons in the mode interact even more strongly with the rest of the atoms and therefore induce the next photon even faster. This process accelerates itself, and the total energy initially stored in the atoms is released in a very short and very bright flash. One can show that the pulse length scales as the inverse of the number of atoms, provided that the effects of the propagation of the light pulse through the medium can be neglected, i.e. as long as the diameter of the atomic cloud is much smaller than the wavelength of the resonant cavity mode. By conservation of energy, the maximal intensity scales as NI_0 with N. This increase by a factor N compared with ordinary fluorescence has led to the name "superradiance".

Superradiance was intensively studied both theoretically and experimentally in the 1970s [107, 108, 109, 110], and more or less complicated situations were considered. Interest focused in particular on the macroscopic quantum fluctuations that show up, for example, in the broad delay time distribution of the intensity peak, which can be understood as amplified quantum fluctuations of the initial state of the atoms. In the following I shall consider a particularly simple form, and I shall not derive the corresponding superradiance master equation. Rather, I would like to state and discuss the basic assumptions necessary for the physical understanding of the effect. Readers interested in the details of the derivation are urged to study [107], which could not be surpassed in clarity, anyway. Extensive reviews of superradiance can be found in [110, 111].

4.2.2 Modeling Superradiance

The system of interest in superradiance is the cloud of atoms. The environment consists of the single resonant cavity mode and a continuum of electromagnetic modes outside the cavity. The latter is necessary for dissipation, since the single cavity mode alone can never serve as a heat bath. It would only lead to coherent back and forth oscillation of energy between the atoms and the modes, as in the well known Jaynes–Cummings model [112]. The resonant cavity mode is somewhat singled out, in the sense that it provides the link between the atoms and the continuum of modes outside the cavity. The latter coupling is brought about by photons that leak out of the cavity owing to nonideal mirrors, i.e. mirrors that do not reflect completely. The rate at which a photon can leak out of the cavity will be denoted by κ. The Hamiltonian for the environment consists of a sum over many harmonic oscillators, one oscillator for each mode of the electromagnetic field. The creation operator for the resonant cavity mode will be denoted by $b^\dagger$, the annihilation operator by b.

4.2 Superradiance Damping in Quantum Optics

Any linear operator on the two-dimensional Hilbert space spanned by a two-level atom can be written as a linear combination of unity and the Pauli matrices σ_x, σ_y and σ_z. If we use the two energy eigenstates of two-level atom number i as a basis, its Hamiltonian has the form $(1/2)\hbar\omega_0\sigma_z^{(i)}$ with $\sigma_z^{(i)} = \text{diag}(1, -1)$. Thus, the system Hamiltonian reads

$$H_s = \frac{1}{2}\hbar\omega_0 \sum_{i=1}^{N} \sigma_z^{(i)}. \tag{4.4}$$

Let us assume that the diameter of the atomic cloud is much smaller than the wavelength of the resonant cavity mode. The coupling Hamiltonian of the atoms to the resonant cavity mode then takes the form

$$H_{\text{int}} = \sum_{i=1}^{N} \hbar g(\sigma_-^{(i)} b^\dagger + \sigma_+^{(i)} b), \tag{4.5}$$

where $\sigma_\pm = \sigma_x^{(i)} \pm i\sigma_y^{(i)}$ are the usual atomic ladder operators. An atom can be excited upon absorption of a photon or be deexcited upon emitting one. The assumption of the small diameter of the atomic cloud has simplified (4.5) in as much as the coupling constant g is the same for all atoms. We can therefore introduce a collective observable, the Bloch vector $\boldsymbol{J} = \sum_{i=1}^{N} \boldsymbol{\sigma}^{(i)}$, where $\boldsymbol{\sigma} = (\sigma_x, \sigma_y, \sigma_z)$. Formally, it is an angular momentum with three spatial components J_x, J_y and J_z and absolute value $j(j+1)$, where j can be an integer or half integer, depending on whether the number of atoms is even or odd. The introduction of the Bloch vector simplifies the problem considerably. Instead of N vector operators $\boldsymbol{\sigma}^{(i)}$, we are left with only the three components of $\boldsymbol{J}$. The simplification is possible because the symmetry of the coupling $g_i = g$ restricts the dynamics to the irreducible representation specified by the initial value of J_z, i.e. for full initial excitation of all atoms to $j = N/2$. Instead of having to deal with the huge 2^N-dimensional Hilbert space, we only have to consider a $2j + 1 = N + 1$-dimensional subspace.

In terms of the Bloch vector, the energy of the atoms is given by

$$H_s = \hbar\omega_0 J_z, \tag{4.6}$$

and the interaction Hamiltonian reads

$$H_{\text{int}} = \hbar g(J_- b^\dagger + J_+ b), \tag{4.7}$$

where now $J_\pm = J_x \pm iJ_y$ is the collective ladder operator. For an atom without a permanent electric dipole moment, the dipole operator has only two off-diagonal elements and is therefore proportional to $\sigma_x^{(i)}$. The total polarization of the atomic cloud is then given by J_x.

Note that the coupling of the atoms to the electromagnetic field outside the cavity is in general *not* only via the leaky resonant cavity mode. The continuum of modes also couples directly to each atom, particularly if the cavity is not entirely closed. This coupling is responsible for spontaneous decay of single atoms (in contrast to collective decay via the cavity mode). Such individual "dancing out of the row" of single atoms is very disturbing for a collective effect like superradiance. I shall therefore assume that spontaneous emission happens only very occasionally, with a rate Γ per atom that is much smaller than any other frequency scale in the problem.

Besides the frequency scales ω_0 and κ, the Rabi frequency $g\sqrt{N}$ of the undamped system and the rate $k_B T/\hbar$, related to temperature, play an important role. We shall consider only very low temperatures, such that $k_B T \ll \hbar\omega_0$. This means that we can neglect thermal photons that might come into the cavity from the outside and randomly excite atoms. Concerning the Rabi frequency, we shall assume that it is much smaller than the escape rate κ. Deexcited atoms are then not excited again. We shall see that this leads to an overdamped motion of the Bloch vector, i.e. the Rabi oscillations that one would see without dissipation are completely damped out. Under these assumptions and for weak coupling of the atoms to the cavity mode, the Markovian master equation

$$\frac{d}{dt}\rho(t) = \gamma([J_-, \rho(t)J_+] + [J_-\rho(t), J_+]) \tag{4.8}$$

was derived in [107] for the reduced density matrix $\rho(t)$. The rate γ is given by $\gamma = g^2/\kappa$. The master equation is of the so-called Lindblad type, the most general type possible if one requires the Markov property, conservation of positivity, and initial decoupling between the system and bath [113].

In spite of its limitations and of all the assumption that have been made in its derivation, (4.8) has been well confirmed in experiments by Haroche and coworkers [108, 114] and by Feld and coworkers [115]. At the time of the experiments, the most interesting physical aspect of superradiance was the fact that initial quantum fluctuations (caused by spontaneous emission from one or a few first atoms) are amplified and lead to fluctuations on a macroscopic scale. For example, the fluctuations of the delay time of the light pulse after the excitation of the atoms are comparable to the average delay time [116]. The measured statistics were in good agreement with theoretical predictions based on (4.8).

For us, the master equation will provide a simple yet physical form of dissipation that we shall use throughout this book as a primary example. It can easily be combined with the unitary quantum map introduced in the previous chapter, that of the kicked top. But before we do so, let us first learn a bit more about the dissipative process itself.

4.2.3 Classical Behavior

To get a feeling for the physics hidden in (4.8), let us look at its *classical* limit. The classical equations of motion can be determined by extracting equations for the expectation values of J_z and $J_\pm$, parameterizing them as in the case of the kicked top as $\langle J_z \rangle = J\cos\theta$ and $\langle J_\pm \rangle = J\sin\theta\, e^{\pm i\phi}$, and factorizing all operator products (e.g. $\langle J_+ J_- \rangle \to \langle J_+ \rangle \langle J_- \rangle$). One then finds that the angular momentum behaves classically, like an overdamped pendulum [60];

$$\dot\phi = 0, \qquad \dot\theta = 2J\gamma \sin\theta. \tag{4.9}$$

The latter equation reveals the classical damping rate as $2J\gamma$. The solution of (4.9) is easily found by a simple integration. In terms of the dimensionless time $\tau = 2J\kappa t$, i.e. the time in units of the classical timescale, and the phase-space variables $\mu = \cos\theta$ and ϕ, we find

$$\tau = \frac{1}{2} \ln \frac{[1-\mu(\tau)][1+\mu(0)]}{[1-\mu(0)][1+\mu(\tau)]}, \qquad \phi(\tau) = \phi(0). \tag{4.10}$$

The Bloch vector moves down towards the south pole $\mu = -1$ of the Bloch sphere on a great circle $\phi = const.$, accelerating on the northern hemisphere and decelerating again on the southern hemisphere, as is evident from rewriting (4.10) as $\tan[\theta(\tau)/2] = e^\tau \tan[\theta(0)/2]$.

For future reference, I rewrite the master equation in terms of the dimensionless time as

$$\frac{\mathrm{d}}{\mathrm{d}\tau}\rho(\tau) = \frac{1}{J}\{[J_-, \rho(\tau) J_+] + [J_- \rho(\tau), J_+]\} \equiv \Lambda\rho(\tau). \tag{4.11}$$

The generator Λ defined in the above equation will be useful for a compact formal representation of the propagator.

4.3 The Short-Time Propagator

In the next section (Sect. 4.4) I shall present a fairly general method for obtaining the propagator of the density matrix for a Markovian master equation like (4.11). The method is valid for times $\tau \gtrsim 1/J$, and this is the regime I shall focus on for the most part of what follows. When discussing decoherence in the next chapter, however, we shall be interested in the opposite regime of very short times. I therefore take the opportunity to construct in this section the short-time propagator. This should also be useful if the semiclassical methods of Chap. 7 are to be generalized to very weak damping.

Let us start by writing the master equation in the $|j,m\rangle$ basis. Denoting the density matrix elements by

$$\langle j, m+k | \rho(\tau) | j, m-k \rangle = \rho_m(k, \tau), \tag{4.12}$$

we obtain

$$J\frac{\mathrm{d}\rho_m(k,\tau)}{\mathrm{d}\tau} = \sqrt{g_{m+k+1} g_{m-k+1}}\, \rho_{m+1}(k,\tau) - (g_m - k^2)\rho_m(k,\tau), \tag{4.13}$$

where g_m denotes the "rate function"

$$g_m = j(j+1) - m(m-1). \tag{4.14}$$

The master equation (4.11) does not couple density matrix elements with different skewness k. In particular, the probabilities ($k = 0$) can be solved for independently of the off-diagonal matrix elements, the so-called coherences ($k \neq 0$). The particular solution $\rho_m(k, \tau)$ satisfying the initial condition $\rho_m(k, \tau = 0) = \delta_{mn}$ for a certain n is called the dissipative propagator and denoted by $D_{mn}(k, \tau)$. Owing to the conservation of skewness, k is just a parameter. The solution for an arbitrary initial density matrix is then $\rho_m(k, \tau) = \sum_{n=-j}^{j} D_{mn}(k, \tau) \rho_n(k, 0)$. Formally, the propagator is given with the help of the generator Λ, by

$$D = \exp(\Lambda \tau). \tag{4.15}$$

The further analysis proceeds via the Laplace image

$$\mathcal{D}_{mn}(k, z) = \int_0^\infty e^{-z\tau} D_{mn}(k, \tau) \, d\tau \tag{4.16}$$

of the propagator. Laplace-transforming (4.13), one is lead to a recurrence relation for the $\mathcal{D}_{mn}(k, z)$, with the easily found solution [107, 117]

$$\mathcal{D}_{mn}(k, z) = \frac{1}{\sqrt{g_{m-k} g_{m+k}}} \prod_{l=m}^{n} \frac{\sqrt{g_{l-k} g_{l+k}}}{z + g_l - k^2}. \tag{4.17}$$

To obtain the dissipative propagator itself we have to invert the Laplace transform. With the help of the quantity

$$Q_{mn} = \prod_{l=m+1}^{n} g_l = \frac{(j+n)!(j-m)!}{(j+m)!(j-n)!}, \tag{4.18}$$

the propagator takes the form

$$D_{mn}(k, \tau) = \frac{\sqrt{Q_{m-k,n-k} Q_{m+k,n+k}}}{2\pi i} \int_{b-i\infty}^{b+i\infty} dv \, e^{\tau v / J} \prod_{l=m}^{n} \frac{1}{v + g_l - k^2}, \tag{4.19}$$

where b should be larger than the largest pole in the denominator. The indices are restricted to $m \leq n$. Otherwise, $D_{mn}(k, \tau) = 0$, since the probabilities and coherences only flow downwards on the J_z ladder.

An unexpected identity connecting the propagators for the diagonal and off-diagonal elements of the density matrix follows immediately from (4.19),

$$D_{mn}(k, \tau) = D_{mn}(0, \tau) \frac{\sqrt{Q_{m-k,n-k} Q_{m+k,n+k}}}{Q_{mn}} e^{k^2 \tau / J}. \tag{4.20}$$

For the proof it is sufficient to shift the integration variable in (4.19) to $\bar{v} = v - k^2$.

Equation (4.19) is an exact integral representation of the propagator. Unfortunately, the integrand contains in general very many poles (most of them degenerate), so that a straightforward analytical back transformation

4.3 The Short-Time Propagator

is very clumsy. However, for very small times τ the structure of the Laplace image can be very much simplified and an analytical inversion is possible. To explain the essence of the approximation, let me give a simple example. Consider a Laplace image function with two simple poles $\mathcal{V}(z) = (z - c - d)^{-1}(z - c + d)^{-1}$ and its original function $V(t) = e^{ct} d^{-1} \sinh td$. As long as $td \ll 1$ the hyperbolic sine can be replaced by its argument, such that $V(t) \approx te^{ct}$. We have thus in effect replaced the closely spaced poles of the Laplace image by a single second-order pole; that replacement is obviously justified for sufficiently small times.

Let me employ this observation for the Laplace representation of the propagator (4.19) for the probabilities. With the new integration variable $x = \tau v/J$, we obtain

$$D_{mn}(0,\tau) = Q_{nm} \left(\frac{\tau}{J}\right)^{n-m} \frac{1}{2\pi i} \int_{b-i\infty}^{b+i\infty} \frac{e^x dx}{\prod_{l=m}^{n}(x + g_l \tau/J)}. \tag{4.21}$$

The length of the interval on which the poles of the integrand now lie is proportional to τ;

$$|g_m - g_n| \frac{\tau}{J} = \frac{|m+n-1|}{J}(n-m)\tau. \tag{4.22}$$

If that length is much smaller than unity, the poles of the integrand of (4.21) are nearly degenerate, and that proximity enables us to replace the product in the denominator by the $(n-m)$th power of the average factor $x + \bar{g}\tau/J$, where $\bar{g} \equiv g_{(m+n)/2} = J^2 - [(n+m-1)/2]^2$. The integral is then easily calculated and yields the small-time asymptotic approximation of the dissipative propagator,

$$D_{mn}(0,\tau) = \frac{Q_{mn}}{(n-m)!} \left(\frac{\tau}{J}\right)^{n-m} \exp\left\{-\frac{\tau}{J}\left[J^2 - \left(\frac{n+m-1}{2}\right)^2\right]\right\}. \tag{4.23}$$

The propagator for general k follows from (4.20);

$$D_{mn}(k,\tau) = \frac{\sqrt{Q_{m-k,n-k}Q_{m+k,n+k}}}{(n-m)!} \left(\frac{\tau}{J}\right)^{n-m}$$
$$\times \exp\left\{-\frac{\tau}{J}\left[j^2 - \left(\frac{n+m-1}{2}\right)^2\right]\right\}. \tag{4.24}$$

According to the condition on the near-degeneracy of the poles (4.22), the validity of (4.22) is limited to times $\tau \ll J/(|m+n-1|(n-m))$. This time is of the order of $1/J$, unless m and n or m and $-n$ are very close together, in which case the result can hold even up to times $\tau \sim 1$.

4.4 The Semiclassical Propagator

In this section I present a rather general method for the solution of master equations of the form (4.11). The only properties needed are the Markovian property and the small factor $1/J$. We shall observe that in the limit of small $1/J$ (i.e. for a large number N of atoms) the master equation becomes a finite-difference equation with a small step, amenable to solution by an approximation of the WKB type. The propagator solution thus obtained takes the form of a Van Vleck propagator involving the action of a certain classical Hamiltonian system with one degree of freedom. But let me show all of this step by step.

4.4.1 Finite-Difference Equation

In the limit of large J it is convenient to use as the independent variable the momentum $\mu = m/J$ defined in Chap. 3 instead of m. I also define its increment Δ as

$$\Delta = J^{-1}. \tag{4.25}$$

In the classical limit μ becomes continuous in the range $-1\ldots 1$. In our semiclassical perspective μ remains discrete but neighboring values are separated by Δ. In the following I shall derive the semiclassical formalism first for the densities ($k = 0$). The propagator for the coherences can always be obtained from (4.20). This will be discussed in Sect. 4.4.7. To simplify the notation, the skewness index k will be dropped till then, and I write the density matrix as $\rho(\mu, \tau) \equiv \rho_m(k = 0, \tau)$.

Expressed in terms of μ and τ, the master equation (4.13) for the densities becomes a finite-difference equation,

$$\frac{\partial \rho(\mu, \tau)}{\partial \tau} = J\left[g_-(\mu, \Delta)\rho(\mu + \Delta, \tau) - g_+(\mu, \Delta)\rho(\mu, \tau)\right], \tag{4.26}$$

$$g_\pm(\mu, \Delta) = \left(1 - \mu^2 \pm \mu\Delta - \frac{\Delta^2}{4}\right). \tag{4.27}$$

4.4.2 WKB Ansatz

The WKB formalism for finite-difference equations with a small step is well established. The general theory has been worked out mostly by Maslov [118]. The WKB method for ordinary second-order difference equations has been extensively used to study the eigenvalues of large tridiagonal matrices occurring in the theory of Rydberg atoms in external fields [94]. Closer to our topic, the leading (exponential) term in the semiclassical solution of master equations of the type (4.27) was obtained in [119]. I follow the same lines but go a step further by also establishing the preexponential factor. We shall see later that the prefactor is indeed the decisive part of the propagator for most applications.

Let us look for a solution of (4.27) in a form reminiscent of the WKB wave function,

$$\rho(\mu,\tau) = A(\mu,\tau)\,e^{JR(\mu,\tau)}. \tag{4.28}$$

Here the prefactor A and the "action" R are smooth functions satisfying the initial conditions

$$R(\mu,0) = R_0(\mu), \qquad A(\mu,0) = A_0(\mu). \tag{4.29}$$

The WKB ansatz for a Schrödinger equation would contain the imaginary unit in the exponent. In (4.28) it is absent, since (4.27) is also real. Owing to the presence of the large parameter J, even modest changes of R_0 are reflected in wild fluctuations of $\rho(\mu,0)$; the ansatz therefore does not limit our discussion to smooth probability distributions.

Assuming the function $R(\mu,\tau)$ to be independent of J does not mean any loss of generality, since the prefactor $A(\mu,\tau)$ may pick up all dependence on J. I represent the latter by an expansion in powers of $\Delta = J^{-1}$,

$$A(\mu,\tau) = A^{(0)}(\mu,\tau) + A^{(1)}(\mu,\tau)\Delta + A^{(2)}\Delta^2 + \ldots. \tag{4.30}$$

The master equation (4.27) then allows one to determine $R, A^{(0)}, \ldots$ recursively. We shall need the equations for the action and the zero-order prefactor,

$$\frac{\partial R}{\partial \tau} + (1-\mu^2)\left(1 - e^{\frac{\partial R}{\partial \mu}}\right) = 0, \tag{4.31}$$

$$\left(\frac{\partial}{\partial \tau} - e^{\partial R/\partial \mu}(1-\mu^2)\frac{\partial}{\partial \mu}\right)\ln A^{(0)} = e^{\partial R/\partial \mu}\left((1-\mu^2)\frac{1}{2}\frac{\partial^2 R}{\partial \mu^2} - \mu\right) - \mu. \tag{4.32}$$

I shall neglect all higher-order corrections to the zero-order prefactor.

4.4.3 Hamiltonian Dynamics

We may consider (4.31) as the Hamilton–Jacobi equation for a classical system with one degree of freedom and the Hamiltonian

$$H(\mu,p) = (1-\mu^2)(1-e^p). \tag{4.33}$$

Note that this Hamiltonian lives in a different phase space than does the original dissipative dynamics. There we had defined the momentum as μ and the canonical coordinate as ϕ. Here μ plays the role of the canonical coordinate, and $p = \partial R/\partial \mu$ the role of momentum. The canonical equations of motion $\dot{\mu} = \partial H/\partial p = -(1-\mu^2)e^p$, $\dot{p} = -\partial H/\partial \mu = 2\mu(1-e^p)$ are easily integrated. They result in the "Hamiltonian" trajectories

$$\tau = \frac{1}{2a} \ln \frac{(a+\nu)(a-\mu)}{(a-\nu)(a+\mu)}, \tag{4.34}$$

$$p = \ln \frac{a^2 - \mu^2}{1 - \mu^2}, \tag{4.35}$$

where $\mu = \mu(\tau)$, $\nu = \mu(0)$. The name "Hamiltonian" is meant to distinguish these solutions from the classical trajectories of the overdamped pendulum (4.10). The second integration constant, a, determines the "energy" $\tilde{E} = H(\mu, p)$ through

$$a \equiv \sqrt{1 - \tilde{E}}. \tag{4.36}$$

The tilde is meant to remind us that the energy here is the energy of the underlying Hamiltonian system, not the energy of the physical system, which of course decreases owing to the dissipation. Rather remarkably, the Hamiltonian trajectory (4.34) coincides with the saddle-point equation encountered in [117] when examining the asymptotics of the Laplace representation of the propagator. For later reference I note the nonnegative "speed",

$$\dot{\mu} = -(a^2 - \mu^2). \tag{4.37}$$

A special class of Hamiltonian trajectories has zero initial momentum, $p(\tau = 0) = 0$, and therefore vanishing energy, $\tilde{E} = 0$, and $a = 1$. These are just the classical trajectories of the classical overdamped pendulum (4.10),

$$\tau = \frac{1}{2} \ln \frac{(1+\nu)(1-\mu)}{(1-\nu)(1+\mu)}, \quad p(t) = 0. \tag{4.38}$$

Their canonical momentum is conserved with value zero. Since p is the logarithmic derivative of the density profile, this means that the maximum of the distribution travels on a classical trajectory.

The semiclassical quantum effects which our Hamiltonian dynamics imparts to the spin through the WKB ansatz (4.28) may be seen in the existence of the Hamiltonian trajectories (4.34) with $a \neq 1$, not included in the special class (4.38).

4.4.4 Solution of the Hamilton–Jacobi Equation

The familiar relation between canonical momentum and action,

$$p = \frac{\partial R(\mu, \tau)}{\partial \mu}, \tag{4.39}$$

implies $p_0(\nu) = \partial R_0(\nu)/\partial \nu$ at the initial moment $\tau = 0$. Since R_0 is fixed by the initial density distribution, this latter equation uniquely associates an initial momentum with the initial coordinate ν. One and only one Hamiltonian trajectory $\mu(\tau; \nu, a)$ thus passes at $\tau = 0$ through the initial coordinate ν, provided of course that we consider the initial probabilities as imposed. Conversely, we can find the initial coordinate $\nu = \nu(\mu, \tau)$ from which the current coordinate μ is reached at time τ along the unique Hamiltonian trajectory.

The action $R(\mu, \tau)$ can be obtained by integration along the trajectory just discussed;

$$R(\mu, \tau) = \left[R_0(\nu) + \int_\nu^\mu p\, d\mu - \tilde{E}\tau \right]_{\nu=\nu(\mu,\tau)}. \tag{4.40}$$

The explicit form of the Hamiltonian trajectories (4.35) allows us to evaluate the integral. The resulting action can be expressed in terms of the auxiliary functions

$$\sigma(a, \mu, \nu) = (\nu + a)\ln(\nu + a) - (\mu + a)\ln(\mu + a)$$
$$- (a - \nu)\ln(a - \nu) + (a - \mu)\ln(a - \mu), \tag{4.41}$$
$$R(\mu, \nu, \tau) = \left[\sigma(1, \mu, \nu) - \sigma(a, \mu, \nu) + \tau(a^2 - 1) \right]_{a=a(\mu,\nu,\tau)} \tag{4.42}$$

as

$$R(\mu, \tau) = \left[R_0(\nu) + R(\mu, \nu, \tau) \right]_{\nu=\nu(\mu,\tau)}. \tag{4.43}$$

In the definition (4.42) of the function $R(\mu, \nu, \tau)$ the parameter a must, as indicated above, be read as a function of the initial and final values of the coordinate since these are at present considered as defining a Hamiltonian trajectory. We may interpret the function $R(\mu, \nu, \tau)$ as the action accumulated along the Hamiltonian trajectory in question. This function should not be confused with $R(\mu, \tau)$; I shall distinguish the two of them by explicitly giving all arguments. The derivatives with respect to μ and ν give the final and initial momenta,

$$\frac{\partial R(\mu, \nu, \tau)}{\partial \mu} = \ln \frac{a^2 - \mu^2}{1 - \mu^2} = p, \tag{4.44}$$

$$\frac{\partial R(\mu, \nu, \tau)}{\partial \nu} = -\ln \frac{a^2 - \nu^2}{1 - \nu^2} = -p_0. \tag{4.45}$$

4.4.5 WKB Prefactor

The expression (4.32) for the prefactor can be simplified using the notion of the full time derivative of a function $f(\mu, \tau)$ along the Hamiltonian trajectory $\mu(\nu, \tau)$,

$$\frac{df(\mu, \tau)}{d\tau} = \left. \frac{\partial f(\mu(\tau, \nu), \tau)}{\partial \tau} \right|_\nu = \left. \frac{\partial f}{\partial \tau} \right|_\mu + \dot{\mu} \left. \frac{\partial f}{\partial \mu} \right|_\tau,$$

since the left-hand side in (4.32) is just the full time derivative dA/dt (see (4.39)). By introducing the Jacobian

$$Y(\tau; \nu, a) = \frac{\partial \mu(\tau, \nu)}{\partial \nu}, \qquad Y(0, \nu) = 1, \tag{4.46}$$

and a new exponent $E(\mu, \tau)$, we can rewrite the prefactor as

$$A = \frac{e^{E(\mu, \tau)}}{\sqrt{Y}}. \tag{4.47}$$

The full time derivative of Y can be transformed to

$$\frac{dY}{d\tau} = \frac{\partial^2 \mu}{\partial \tau \partial \nu} = \frac{\partial}{\partial \nu}\dot{\mu} = \frac{\partial}{\partial \nu}\frac{\partial H(\mu,p)}{\partial p} = \frac{\partial \mu}{\partial \nu}\left(\frac{\partial^2 H}{\partial \mu \partial p} + \frac{\partial p}{\partial \mu}\frac{\partial^2 H}{\partial p^2}\right)$$

$$= Y\left(\frac{\partial^2 H}{\partial \mu \partial p} + \frac{\partial^2 R(\mu,\tau)}{\partial \mu^2}\frac{\partial^2 H}{\partial p^2}\right)$$

$$= Y \exp\left(\frac{\partial R(\mu,\tau)}{\partial \mu}\right)\left(2\mu - (1-\mu^2)\frac{\partial^2 R(\mu,\tau)}{\partial \mu^2}\right). \qquad (4.48)$$

So equipped, we find the simple evolution equation $dE/d\tau = -\mu$ for the function $E(\mu,\tau)$. It can be integrated along the trajectory, giving

$$E(\mu,\tau) = -\int_0^\tau \mu\, d\tau + \ln A(\nu, 0) = -\frac{1}{2}\ln\frac{a^2 - \mu^2}{a^2 - \nu^2} + \ln A(\nu, 0). \qquad (4.49)$$

We thus arrive at the asymptotic solution of the Cauchy problem for our master equation with the initial condition (4.29),

$$\rho(\mu,\tau) = \frac{1}{\sqrt{\partial \mu(\nu,\tau)/\partial \nu}}\sqrt{\frac{a^2-\nu^2}{a^2-\mu^2}}e^{JR(\mu,\nu,\tau)}\rho(\nu,0), \qquad (4.50)$$

where ν, a mean functions of μ and τ as explained above.

4.4.6 The Dissipative Van Vleck Propagator

The dissipative propagator establishes a linear relation between the initial and final density matrix elements. In the limit of large J the sum in this relation can be replaced by an integral; using the classical variables μ, ν as arguments, it can be written as (in the case $k = 0$)

$$\rho(\mu,\tau) = \int_{-1}^{1} d\nu D(\mu,\nu,\tau)\rho(\nu,0), \qquad (4.51)$$

where the function $D(\mu,\nu,\tau)$ is related to the matrix $D_{mn}(\tau)$ by $D(\mu,\nu,\tau) = JD_{mn}(\tau)|_{m=J\mu,\, n=J\nu}$.

To obtain the propagator one has to solve the master equation with a δ peak as the initial density distribution. Strictly speaking, such an initial condition does not fall into the class (4.28), so that our solution of the Cauchy problem (4.50) is not directly applicable. It is easy, however, to extract the propagator out of (4.50) in a slightly roundabout way. The semiclassical solution of the dissipative problem in the form (4.28) points to an analogy between our master equation for the densities and a Schrödinger equation in imaginary time. In the spirit of that analogy, we may consider the function $D(\mu,\nu,\tau)$ in (4.51) as the Van Vleck propagator [87], which must have the structure

$$D(\mu,\nu,\tau) = B(\mu,\nu,\tau)e^{JR(\mu,\nu,\tau)}, \qquad (4.52)$$

with $R(\mu,\nu,\tau)$ being the action accumulated along the trajectory.

Our task is to establish the prefactor B. To do so let us substitute the initial density (4.28) and the semiclassical propagator in the form (4.52) into (4.51) and perform the integration by saddle-point approximation. The maximum $\nu^* = \nu^*(\mu, \tau)$ of the exponent defines the Hamiltonian trajectories. The saddle-point integration thus gives

$$\rho(\mu, \tau) = B(\mu, \nu, \tau)$$
$$\times \sqrt{\frac{2\pi}{J}} \left(-\frac{\partial^2 [R(\mu, \nu, \tau) + R_0(\nu)]}{\partial \nu^2} \right)^{-1/2} e^{JR(\mu,\nu,\tau)} \rho(\nu, 0), \qquad (4.53)$$

where $\nu^*(\mu, \tau)$ should be substituted for ν. Comparing with (4.50), we find the prefactor;

$$B = \sqrt{\frac{J}{2\pi}} \left(-\frac{\partial^2}{\partial \nu^2} [R(\mu, \nu, \tau) + R_0(\nu)] \right)^{1/2} \frac{\sqrt{(a^2 - \nu^2)/(a^2 - \mu^2)}}{\sqrt{\partial_\nu \mu(\tau, \nu)}}. \qquad (4.54)$$

A simpler form results if we differentiate (4.43) with respect to ν and μ;

$$\frac{\partial \mu}{\partial \nu} = \frac{-\partial_\nu^2 [R(\mu, \nu, \tau) + R_0(\nu)]}{\partial_\mu \partial_\nu R(\mu, \nu, \tau)}. \qquad (4.55)$$

For the propagator, we thus find

$$D_{mn}(\tau) = \frac{1}{\sqrt{2\pi J}} \sqrt{\partial_\nu \partial_\mu R(\mu, \nu, \tau)} \sqrt{\left(\frac{\partial \nu}{\partial \mu}\right)_{\tilde{E}}} e^{JR(\mu,\nu,\tau)}. \qquad (4.56)$$

Compared with the Van Vleck propagator for unitary quantum maps (3.11), the preexponential factor in this expression contains an additional square root factor, the origin of which can be traced to the difference in the normalization conditions for wave functions and density matrices. It is system specific in as much as $\tilde{E}$ is the conserved energy of the fictitious Hamiltonian system. However, on the classical trajectory ($\tilde{E} = 0$) this factor is just the square root of the classical Jacobian for the inverted classical trajectory (4.38). Both square roots in (4.56) give rise to the same factor in the classical limit and combine to give the Jacobian to the power one, as is necessary to guarantee probability conservation. The explicit form of the additional prefactor for the present problem is given by $\sqrt{[\partial_\mu \nu(\mu, \tau)]_{\tilde{E}}} = \sqrt{(a^2 - \nu^2)/(a^2 - \mu^2)}$.

4.4.7 Propagation of Coherences

No separate investigation of the coherence propagator $D_{mn}(k, \tau)$ is necessary, because of the identity (4.20). If desired, the propagator can be approximated semiclassically by replacing factorials via Stirling's formula. It is instructive, however, to consider the changes in our Hamilton–Jacobi formalism necessitated by nonzero k. The new quantum number k, whose range goes to infinity when $j \to \infty$, is accompanied by a macroscopic variable

$$\eta = \frac{k}{J}. \tag{4.57}$$

The master equation (4.13), written with μ, ν as arguments and the convention $\rho(\mu, \eta, \tau) = \rho_m(k, \tau)$, reads

$$J\frac{\partial \rho(\mu, \eta, \tau)}{\partial \tau}$$
$$= J^2 \sqrt{[1 - (\mu + \eta)^2 - (\mu + \eta)\Delta][1 - (\mu - \eta)^2 - (\mu - \eta)\Delta]} \, \rho(\mu + \Delta, \eta, \tau)$$
$$- J^2 \left(1 - \mu^2 - \eta^2 + \mu\Delta\right) \rho(\mu, \eta, \tau) + \mathcal{O}(\Delta^2).$$

A Hamilton–Jacobi ansatz

$$\rho(\mu, \eta, \tau) = A(\mu, \eta, \tau)\, e^{JR(\mu, \eta, \tau)} \tag{4.58}$$

entails a chain of differential equations for the "action" R and for the terms in the expansion of the amplitude A in powers of Δ. I shall examine only the Hamilton–Jacobi equation

$$\frac{\partial R}{\partial \tau} + G(\mu) - F(\mu) \exp\left(\frac{\partial R}{\partial \mu}\right) = 0, \tag{4.59}$$

where F and G denote the auxiliary functions

$$F(\mu) = \sqrt{[1 - (\mu + \eta)^2][1 - (\mu - \eta)^2]}, \quad G(\mu) = 1 - \mu^2 - \eta^2. \tag{4.60}$$

The previous Hamiltonian becomes extended to

$$H(\mu, p) = G(\mu) - F(\mu)e^p. \tag{4.61}$$

Denoting once more the conserved value of H by $\tilde{E}$ and introducing the constant a for finite η by the relation

$$a = \sqrt{1 - \tilde{E} - \eta^2}, \tag{4.62}$$

we obtain the canonical equation for the coordinate,

$$\dot{\mu} = -F e^p = a^2 - \mu^2. \tag{4.63}$$

This coincides with (4.37), and its integration leads to exactly the same trajectories $\mu = \mu(\nu, \tau)$ (4.34) as for the densities. The characteristic lines for the propagation of the coherences are the same as the ones for the propagation of the probability. The rest of the calculation follows exactly the same steps as for $k = 0$ [15]. Let me give the result right away for the propagator $D_{mn}(k, \tau) = (1/J)D(\mu, \nu; \eta; \tau)$:

$$D(\mu, \nu; \eta; \tau) = B(\mu, \nu; \eta; \tau) \exp[JR(\mu, \nu; \eta; \tau)], \tag{4.64}$$

$$R(\mu, \nu; \eta; \tau) = \frac{1}{2}[\xi(1, \nu - \eta) - \xi(1, \mu - \eta) + \xi(1, \nu + \eta) - \xi(1, \mu + \eta)]$$
$$- \xi(a, \nu) + \xi(a, \mu) + \tau(a^2 - 1 + \eta^2), \tag{4.65}$$

$$\xi(x, y) \equiv (x + y)\ln(x + y) - (x - y)\ln(x - y), \tag{4.66}$$

$$B(\mu, \nu; \eta; \tau) = \sqrt{\frac{J}{2\pi}} \sqrt{\left.\frac{\partial \nu}{\partial \mu}\right|_{\tilde{E}}} \sqrt{\partial_\mu \partial_\nu R(\mu, \nu; \eta; \tau)}. \tag{4.67}$$

The second square root is given again by $\sqrt{(a^2 - \nu^2)/(a^2 - \mu^2)}$.

With (4.64)–(4.67), we have the full semiclassical propagator for superradiance damping in our hands. It will be used many times in Chap. 6. Before concluding this chapter with a numerical check of this propagator and a comment on its limitations, let me therefore point out several features of the action R which will be of importance for the later semiclassical analyses.

4.4.8 General Properties of the Action R

The features in question are the following:

- For $\eta = 0$ we obtain $a = 1$ from $\partial_\mu R = 0$, and thus the classical equation of motion (4.10). If this equation is fulfilled, $\partial_\nu R = 0$ holds as well.
- For $\eta = 0$ and μ, ν connected by the classical trajectory (i.e. $a = 1$), R is strictly zero. This actually holds beyond the semiclassical approximation and can be traced back to conservation of probability by the master equation for $k = 0$. To show this I write (4.11) in the J_z basis and look at the part with vanishing skewness, i.e. the probabilities $p_m = \langle m|\rho|m\rangle$. We obtain a set of equations

$$\frac{\mathrm{d}}{\mathrm{d}\tau} p_m = g_{m+1} p_{m+1} - g_m p_m , \qquad (4.68)$$

where the specific form of the coefficients g_m is of no further concern. The important point is, rather, that the *same* function g_m appears twice. This is sufficient and necessary for the conservation of probability, $\mathrm{tr}\,\rho = \sum_{m=-j}^{j} p_m = 1$. On the other hand, if we had coefficients f_m and g_m, i.e. $\dot{p}_m = g_{m+1} p_{m+1} - f_m p_m$, we would obtain the action R on the classical trajectory as $JR = \sum_{l=m}^{n}(\ln g_l - \ln f_l)$, as one can easily verify by writing down the exact Laplace image of D following the lines of [117]. Thus, the action is zero iff probability is conserved.
- R is an even function of η and has always a *maximum* at $\eta = 0$.

4.4.9 Numerical Verification

To demonstrate the accuracy of the dissipative Van Vleck propagator, I compare in Fig. 4.1 the time dependence of the probability distribution as obtained by the WKB method with the numerically "exact" solution.

The latter was obtained by numerically inverting the exact Laplace image (4.17). Owing to the many poles, this is done most conveniently by actually performing the integration. One chooses an appropriate integration path in the complex plane, namely a path of constant phase, which can easily be found approximately.

The data in Fig. 4.1 represent a case where the system is initially in a pure coherent state of the angular momentum, $|\gamma\rangle$. A coherent state is a smooth wave packet with the minimum possible uncertainty (see Sect. 5.3.1 for a precise definition). The complex label γ determines the direction of

48   4. Dissipation in Quantum Mechanics

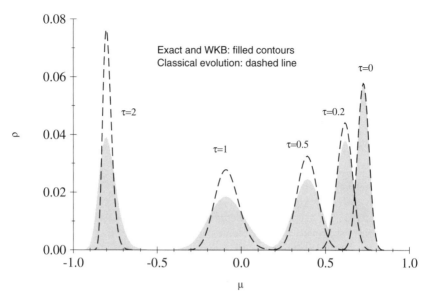

Fig. 4.1. The time evolution of a density profile calculated from the semiclassical propagator ($j = 200$) is practically identical to the exact evolution (*shaded areas* in both cases). The classical propagator (*dashed lines*) reproduces correctly only the position of the maximum, not its height

the mean angular-momentum vector $\langle \gamma | \boldsymbol{J} | \gamma \rangle$ as $\gamma = \tan(\theta/2) e^{i\phi}$. For $j = 200, \gamma = 0.4$ the probability distributions given by the WKB formula coincide with the exact ones to an accuracy in the range 0.4%–1.7% (the accuracy decreases in the later stages of the evolution). The corresponding plots are indistinguishable. The fully classical formula correctly places the probability peaks but grossly underestimates their broadening with time, leading to a 20%–150% error in the width and amplitude. This error does not diminish as j grows!

4.4.10 Limitations of the Approach

Besides giving examples of how well the semiclassical approximation works, let me also say a word about the limitations of the method. The most severe limitation comes from the continuum approximation of the density (or coherence) profiles. This restricts the propagator to times τ that are larger than $1/J$. For smaller times, a density profile initially concentrated in a single level has no time to substantially occupy neighboring levels.

This limitation on τ should be always kept in mind in the subsequent calculations. In particular, it will prohibit us from describing uniformly the whole transition regime from zero to classically finite dissipation, as τ will be a measure of the dissipation strength in the dissipative quantum maps that

I shall introduce in Chap. 6. A uniform propagator that is valid virtually everywhere can also be derived [117]. It is, however, unclear how to write it in terms of classical quantities (classical trajectories, actions, stabilities, etc.), and the uniform propagator is therefore less well adapted for the subsequent semiclassical approximations.

4.5 Summary

We have seen in this chapter a fairly general method by which Markovian master equations with a small parameter can be solved. The solution proceeded in close analogy to the well-known WKB approximation in quantum mechanics and yielded a propagator that could be written in the form of a Van Vleck propagator. Not only the exponential behavior was calculated, but also the prefactor (i.e. the next order in the expansion in the small parameter that controls "classicality"). The prefactor will turn out to be extremely important for the semiclassical methods of Chap. 7. As physical example, superradiance damping was studied in detail.

5. Decoherence

In 1932 Erwin Schrödinger proposed a Gedankenexperiment in which he superimposed a cat in the two states "cat alive" and "cat dead" [120]. Why is such a "burlesque" superposition never encountered in everyday life, even though it should be allowed according to the fundamental superposition principle of quantum mechanics?

A superposition of macroscopically distinct states has become known as a Schrödinger cat state. Today we are aware of several reasons for their absence from our classical experience. Besides problems with the orders of magnitude, and with preparation and detection, the most important reason is decoherence, an effect caused by the interaction with the environment. I am going to discuss decoherence in some detail in the present chapter. After a general introduction I shall describe the only recently discovered possibility of the *absence* of the decoherence for states that profit from a *symmetric* coupling to the environment. As an illustration I shall focus again on superradiance damping. We shall encounter Schrödinger cat states of puzzlingly long lifetime.

5.1 What is Decoherence?

Two wave functions $\psi_1(x)$ and $\psi_2(x)$ are said to be coherent if they have a well-defined phase relation. This means that the relative phase between $\psi_1(x)$ and $\psi_2(x)$ is constant or at least a deterministic function of time, i.e. it does not fluctuate randomly with time. The superposition $\psi(x) = [\psi_1(x) + \psi_2(x)]/\sqrt{2}$ of two coherent wave functions gives rise to interference effects in the probability density $|\psi(x)|^2$. These effects show up in the so-called interference terms $\psi_1(x)\psi_2^*(x) + \psi_2(x)\psi_1^*(x)$, whereas the diagonal terms $|\psi_1(x)|^2$ and $|\psi_2(x)|^2$ correspond to classical probability densities. If the system described by the wave function ψ is coupled to an environment, the relative phase between ψ_1 and ψ_2 will typically fluctuate with time, and the interference terms will rapidly average to zero. Their vanishing is called decoherence.

So in quantum mechanics the coupling of a system to an environment induces not only dissipation of energy, but also decoherence, i.e. the different components of the wave function lose their ability to interfere. In the language of density matrices, this process manifests itself in the decay of the off-diagonal elements of the reduced density matrix ρ. These elements are therefore also termed "coherences". A density matrix corresponding to the pure state $|\psi\rangle = (|\psi_1\rangle + |\psi_2\rangle)/\sqrt{2}$ with $\langle\psi_1|\psi_2\rangle = 0$ has $\rho_{ij} = \langle\psi_i|\rho|\psi_j\rangle = 1/2$ for all $i,j = 1,2$, whereas the off-diagonal elements ρ_{12} and ρ_{21} are zero for a statistical mixture of $|\psi_1\rangle$ and $|\psi_2\rangle$.

Decoherence is believed to be one of the key ingredients in the transition from quantum to classical mechanics [98, 101, 121, 122, 123, 124, 125, 126, 127, 128, 129, 130, 131]. It has been shown in many examples (see T. Dittrich in [42]) that in a macroscopic system the decoherence timescale T_{dec}, i.e. the timescale on which coherence is lost, is typically many orders of magnitude smaller than the classical timescale T_{class} on which probabilities evolve. If a classical distance can be attributed to the components in a superposition, the decoherence rate typically scales as a positive power of this distance in units of a microscopic scale. Let me give two examples.

The reduced density matrix of a weakly damped harmonic oscillator with an ohmic heat bath evolves according to the master equation

$$\dot{\rho} = \kappa\left([a, \rho a^{\dagger}] + [a\rho, a^{\dagger}]\right) \equiv \Lambda_{\text{osc}}\rho, \tag{5.1}$$

where $[a, a^{\dagger}] = 1$; this equation is closely analogous to the superradiance master equation (4.11) and in fact underlies the latter as the dynamics of the resonator mode [107]. Coherent states of the oscillator are eigenstates of a, $a|\alpha\rangle = \alpha|\alpha\rangle$, with the complex eigenvalue α related to the mean number of quanta of excitation by $\langle\alpha|a^{\dagger}a|\alpha\rangle = |\alpha|^2$. Two coherent states $|\alpha\rangle$ and $|\beta\rangle$ have a scalar product $\langle\beta|\alpha\rangle = \exp[\beta^*\alpha - \frac{1}{2}|\alpha|^2 - \frac{1}{2}|\beta|^2]$. It is not difficult to show [124] that the quantum coherence between two such states decays according to $e^{\Lambda_{\text{osc}}t}|\alpha\rangle\langle\beta| = \langle\beta|\alpha\rangle^{(1-e^{-2\kappa t})}|\alpha e^{-\kappa t}\rangle\langle\beta e^{-\kappa t}|$. The damping rate κ is a classical damping rate as it determines the decay of the probabilities, $\langle\alpha|\rho|\alpha\rangle$. For times much smaller than $1/\kappa$ we can expand the exponential and find, since $\langle\alpha|\beta\rangle = \exp(-|\beta-\alpha|^2/2)$ that $e^{\Lambda_{\text{osc}}t}|\alpha\rangle\langle\beta| \simeq \exp(-|\beta-\alpha|^2\kappa t)|\alpha e^{-\kappa t}\rangle\langle\beta e^{-\kappa t}|$. The quantity $|\beta - \alpha|^2$ indicates the distance between the two coherent states in phase space if position and momentum are measured in the microscopic units $\sqrt{\hbar/(m\omega)}$ and $\sqrt{\hbar m\omega}$, respectively. Compared with the damping, the decoherence is accelerated therefore by a factor $|\beta - \alpha|^2$, which is huge if the two states are macroscopically different.

The second example is a free particle of mass m with position x in one dimension interacting with a scalar field $\varphi(q,t)$, which constitutes the environment through $H_{\text{int}} = \epsilon x\, d\varphi/dt$, where ϵ is a constant [125]. Zurek derived a master equation for the reduced density matrix in the position basis of the particle. The coherences $\rho(x, x')$ decay on a timescale

$$T_{\text{dec}} \simeq T_{\text{class}} \frac{\hbar^2}{2mk_{\text{B}}T(x-x')^2} = T_{\text{class}}\left(\frac{\lambda_T}{x-x'}\right)^2$$

if the field is initially in thermal equilibrium at temperature T; here λ_T denotes the thermal de Broglie wavelength, $\lambda_T = \sqrt{\hbar/(2mk_\mathrm{B}T)}$, and k_B is Boltzmann's constant. Owing to the fantastic smallness of this microscopic length scale, decoherence is much much faster than the evolution of the diagonal elements, which takes place on the classical timescale T_class. The huge acceleration factor between the decoherence rate and the classical damping rate has led to the name "accelerated decoherence" [60].

Decoherence is a basis-dependent phenomenon. Obviously, if a reduced density matrix has become diagonal in a given basis, it will contain off-diagonal elements (i.e. "coherences") in another basis. But in which basis does a density matrix become diagonal?

Zurek has shown that the selected basis depends on the form of the coupling to the environment [122], as well as on the relative strengths of the interaction and the system Hamiltonians. The environment selects the basis in which the density matrix becomes diagonal. He termed the process "einselection" and the states forming such a basis "pointer states". The relative strengths of H_int and H_s are determined, according to Zurek, by the density of states of the environment. If the environment has a nonvanishing density of states for energies much larger than the smallest level separation in the system, the interaction Hamiltonian dominates the selection process of the pointer states. They are then eigenstates of a system operator that commutes with the interaction Hamiltonian. So if the coupling between the system and the environment is via position, $H_\mathrm{int} = \sum_k g_k x x_k$ with very many modes k and coupling constants g_k, the pointer states are position eigenstates. If, however, the environment contains only frequencies much smaller than the smallest level spacing in the system, energy eigenstates of the system are selected as pointer states [131]. This explains why atoms are typically found in energy eigenstates, but even fairly small molecules appear localized in position space. An intermediate situation occurs, for example, in Brownian motion. There both the coupling Hamiltonian and the system Hamiltonian are important. The pointer states are localized in phase space, even if the coupling is only via position.

5.2 Symmetry and Longevity: Decoherence-Free Subspaces

It might appear as if accelerated decoherence is an inevitable fact, a fundamental natural law. This is, however, not the case. It is well known by now that certain subspaces of Hilbert space might be completely decoherence-free [122, 132, 133]. Such a situation arises if the coupling to the environment has a certain symmetry, in the sense that the interaction Hamiltonian has degenerate eigenvalues. If $|\psi_1\rangle, |\psi_2\rangle, \ldots, |\psi_n\rangle$ are eigenstates of H_int with the same eigenvalue, then there is no accelerated decoherence in the subspace they span.

The physical principle behind this is very simple. If the system Hamiltonian can be neglected, the states of the system are propagated by $e^{-iH_{\text{int}}t/\hbar}$. States that are eigenstates of H_{int} with the same eigenvalue acquire exactly the same phase factors as a function of time. Therefore the phase coherence between such states remains intact. The dimension of such decoherence-free subspaces can be very large [133, 134], as we shall see in Sect. 5.3.7.

The benefits of symmetric couplings have been known for a long time. For example, it is well known that rotational tunneling of small molecular groups attached to large molecules can be observed up to temperatures much higher than what would correspond to the tunneling frequency. The reason is that the coupling to the environment has exactly the same symmetry as the hindering potential [105, 135, 136, 137, 138, 139, 140, 141]. This is very much in contrast to the ordinary tunnel effect in a linear coordinate x, where decoherence sets in at temperatures comparable to the tunneling frequency. Another way of phrasing the robustness against decoherence in rotational tunneling is to say that single-phonon transitions in the tunneling-split ground state are forbidden owing to selection rules originating from the symmetry of the coupling to the environment. The latter consists here basically of normal vibration modes of the carrier molecule or of the crystal in which it is embedded.

Recently, decoherence-free subspaces have attracted renewed attention in quantum computing. Lidar et al. [132] have examined general Markovian master equations of the Lindblad form

$$\frac{\partial \rho}{\partial t} = -\frac{i}{\hbar}[H_{\text{s}}, \rho] + L_{\text{D}}[\rho], \tag{5.2}$$

$$L_{\text{D}}[\rho] = \frac{1}{2}\sum_{\alpha,\beta=1}^{M} a_{\alpha\beta} L_{\alpha\beta}[\rho], \tag{5.3}$$

$$L_{\alpha\beta}[\rho] = [F_\alpha, \rho F_\beta^\dagger] + [F_\alpha \rho, F_\beta^\dagger], \tag{5.4}$$

where the coefficients $a_{\alpha\beta}$ form a Hermitian matrix. The system operators F_α are known as "coupling agents" [142] or, in the context of quantum computing, as "error generators" [132]. They span an M-dimensional Lie algebra $\mathcal{L}$. A decoherence-free subspace (DFS) is defined as all states ρ with $L_{\text{D}}[\rho] = 0$, since then only the unitary evolution according to the first term in (5.2) remains.

Lidar et al. have shown that the DFS is spanned by degenerate simultaneous eigenstates of all the coupling agents, where

$$F_\alpha|i\rangle = c_\alpha|i\rangle, \quad \text{for all } \alpha. \tag{5.5}$$

A particularly important example is constituted by $c_\alpha = 0$.

A DFS defined by $L_{\text{D}}[\rho] = 0$ is not entirely decoherence-free. The reason is that the unitary system dynamics typically moves the states out of the DFS. But this happens only on a classical timescale, so that accelerated decoherence *is* prevented. Lidar et al. call this "decoherence-free to first order".

Complete absence of decoherence can be achieved if additionally $[H_\mathrm{s}, \rho] = 0$, a situation necessary for applications in quantum computing [134]. Another strategy for preventing decoherence for certain states is the so-called quantum reservoir engineering [143, 144].

In the next section we shall examine DFSs more closely for the case of superradiance damping.

5.3 Decoherence in Superradiance

In this section I shall be mainly concerned with decoherence between so-called angular-momentum coherent states of the Bloch vector in superradiance. Bloch vectors pointing in different directions are examples of macroscopically distinct states, and we would expect that a Schrödinger cat state composed of two such states would lose its phase coherence very rapidly. But we shall see soon that rather spectacular exceptions are possible. Whereas the main part of this section focuses on the $j = N/2$ irreducible representation of $SU(2)$, the last subsection will be devoted to an even larger DFS with parts in all of the $SU(2)$ irreducible representations. But before getting there, let me first define angular-momentum coherent states.

5.3.1 Angular-Momentum Coherent States

The states in angular-momentum Hilbert space that approximate classical states as closely as possible are the so-called angular-momentum coherent states $|\gamma\rangle = |\theta, \phi\rangle$ [145, 146]. They correspond to a classical angular momentum pointing in the direction given by the polar angles θ and ϕ, with the complex label γ given by the stereographic-projection relation $\gamma = \tan(\theta/2)\mathrm{e}^{\mathrm{i}\phi}$. Their uncertainty could not be less, since $\Delta p \Delta q \sim 1/J$, and $1/J$ is the effective $\hbar$ in the problem. Two states with vanishing overlap are macroscopically distinct. In terms of $|jm\rangle$ states, one has the expansion

$$|\gamma\rangle = (1+\gamma\gamma^*)^{-j} \sum_{m=-j}^{j} \gamma^{j-m} \sqrt{\binom{2j}{j-m}} |jm\rangle. \tag{5.6}$$

Coherent states may be more familiar from the case of the harmonic oscillator, where they are eigenstates of the annihilation operator. The compactness of Hilbert space prevents the existence of exact eigenstates of the coupling agent J_-. However, angular-momentum coherent states are approximate eigenstates of J_- in the sense that the angle between $J_-|\gamma\rangle$ and $|\gamma\rangle$ is of the order of $1/\sqrt{j}$. Indeed, if we define this angle α for real γ by $|\langle\gamma|J_-|\gamma\rangle|^2 = \langle\gamma|\gamma\rangle\langle\gamma|J_+J_-|\gamma\rangle\cos^2\alpha$, one can easily show that

$$\cos^2\alpha = \frac{\sin^2\theta}{\sin^2\theta + (2/j)\cos^4(\theta/2)} \tag{5.7}$$

$$= 1 - \frac{1}{j}\frac{1}{2\gamma^2} + \mathcal{O}\left(\frac{1}{j^2}\right) \text{ for } \sin\theta \neq 0. \tag{5.8}$$

Angular-momentum coherent states therefore qualify as pointer states in the limit $j \to \infty$. The corresponding approximate eigenvalues are given by $J_-|\gamma\rangle \simeq j\sin\theta\, e^{-i\phi}|\gamma\rangle$ and immediately reveal a fundamental symmetry: since $\sin\theta_1 = \sin\theta_2$ for $\theta_2 = \pi - \theta_1$, *two different coherent states $|\pi/2-\theta\rangle$ and $|\pi/2+\theta\rangle$ have the same approximate eigenvalue.* We expect, therefore, slow decoherence between coherent states related in this way, and this is what I am going to show in the rest of the section. We shall see that even the small deviations from degeneracy brought about by the fact that the coherent states are only approximate eigenstates of the coupling agent do not give rise to accelerated decoherence.

5.3.2 Schrödinger Cat States

In the following we shall study the decoherence of Schrödinger cat states $|\Phi\rangle$ composed of two angular-momentum coherent states,

$$|\Phi\rangle = c_1|\gamma_1\rangle + c_2|\gamma_2\rangle, \tag{5.9}$$

where c_1 and c_2 are properly normalized but otherwise arbitrary complex coefficients. Such a pure state corresponds to the initial density matrix $\rho(0) = |c_1|^2|\gamma_1\rangle\langle\gamma_1| + c_1 c_2^*|\gamma_1\rangle\langle\gamma_2| + c_1^* c_2|\gamma_2\rangle\langle\gamma_1| + |c_2|^2|\gamma_2\rangle\langle\gamma_2|$. The decoherence process manifests itself in the decay of the off-diagonal elements. Since the evolution equation (4.11) of $\rho(\tau) = \exp(\Lambda\tau)\rho(0)$ is linear, it suffices to discuss the fate of $\exp(\Lambda\tau)(|\gamma\rangle\langle\gamma'|) \equiv \rho(\gamma, \gamma', \tau)$, where γ and γ' may take the values γ_1 and γ_2. The relative weights of the four components $\rho(\gamma, \gamma', \tau)$ can be studied in terms of either of the norms

$$N_1(\gamma, \gamma', \tau) = \operatorname{tr}\rho(\gamma, \gamma', \tau)\rho^\dagger(\gamma, \gamma', \tau), \tag{5.10}$$

$$N_2(\gamma, \gamma', \tau) = \sum_{m_1, m_2 = -j}^{j} |\langle j, m_1|\rho(\gamma, \gamma', \tau)|j, m_2\rangle|. \tag{5.11}$$

The first of these norms is normalized to an initial value of unity, whereas the second should always be considered relative to its value at time $\tau = 0$. Depending on the context, N_1 or N_2 is more convenient from a technical point of view, as we shall see presently. We shall start by calculating the decoherence rate at $\tau = 0$.

5.3.3 Initial Decoherence Rate

At $\tau = 0$, the decoherence rate can be obtained very easily from $dN_1(\tau)/d\tau = \operatorname{tr}[(d\rho/d\tau)\rho^\dagger + (\rho d\rho^\dagger/d\tau)]$ by inserting the master equation (4.13) and observing the action of $J_\pm$ on a coherent state,

$$\langle\theta, \phi|J_\pm|\theta, \phi\rangle = j\sin\theta\, e^{\pm i\phi}. \tag{5.12}$$

We obtain the exact result

$$\left.\frac{dN_1(\tau)}{d\tau}\right|_{\tau=0} = -2J\left[\sin^2\theta_1 + \sin^2\theta_2 - 2\cos(\phi_2-\phi_1)\sin\theta_1\sin\theta_2\right]$$
$$-\left[(1+\cos\theta_1)^2(1+\cos\theta_2)^2\right]. \quad (5.13)$$

Clearly, the first term describes a rate of change on a timescale $\tau \sim 1/J$. Since τ is the time in units of the classical (damping) timescale, we recover the general case of accelerated decoherence. Thus, for a superradiant Schrödinger cat also, decoherence is in general a large factor J faster than the classical damping. However, if $\phi_1 = \phi_2$ and $\sin\theta_1 = \sin\theta_2$, the first term vanishes identically. The second term gives rise only to evolution on a timescale $\tau \sim 1$, i.e. the classical timescale. So, obviously, there is no fast decoherence for a Schrödinger cat in which the two components are identical, as should of course be the case. But since the sine function is symmetric with respect to $\pi/2$, *accelerated decoherence is also absent for Schrödinger cats in which the two components lie on the same great circle $\phi_1 = \phi_2 = \phi$ and are arranged symmetrically about the equator, i.e. $\theta_1 = \pi/2 - \theta$, $\theta_2 = \pi/2 + \theta$.*

5.3.4 Antipodal Cat States

The foregoing result can be confirmed by a special example for which the entire time dependence can be calculated exactly, namely a cat state composed of $|j,j\rangle$ and $|j,-j\rangle$. These states are coherent states as well. The first one corresponds to $\gamma = 0$, the second one to $\gamma = \infty$. The density matrix corresponding to their superposition contains as the only coherences the matrix elements $\rho_0(\pm j, \tau)$ (see Chap. 4 for the $\rho_m(k,\tau)$ notation). Their time dependence can be immediately obtained from (4.19) since only one pole contributes. Alternatively, we can see from the master equation (4.13) that they obey an uncoupled differential equation for $\langle j,j|\rho(\tau)|j,-j\rangle$, $(d/d\tau)\langle j,j|\rho|j,-j\rangle = -\langle j,j|\rho(\tau)|j,-j\rangle$. The norm of the off-diagonal parts therefore decays as

$$N_1(0,\infty,\tau) = N_1(\infty,0,\tau) = e^{-2\tau}, \quad (5.14)$$

i.e. on a classical timescale! The same conclusion can be drawn from the second norm, which yields $N_2(0,\infty,\tau) = N_2(\infty,0,\tau) = e^{-\tau}$. *The polar antipodal cat is therefore definitely long-lived.*

5.3.5 General Result at Finite Times

One might object that (5.13) is only valid at $\tau = 0$, and (5.14) only for antipodal states. A result for finite τ and arbitrary (real) γ_1, γ_2 is called for. Its derivation turns out to be technically rather involved. Let me therefore just present the result and show numerical evidence for its correctness. Readers interested in the details of the calculation are invited to study [142].

For finite times τ with $J\tau \ll 1$, a semiclassical evaluation of the norm $N_2(\tau)$ for $\phi_1 = \phi_2 = 0$ based on the short-time propagator (4.23) leads to

$$\frac{N_2(\tau)}{N_2(0)} = \exp\left(-2J\frac{(\gamma_1-\gamma_2)^2(1-\gamma_1\gamma_2)^2}{[(1+\gamma_1^2)(1+\gamma_2^2)]^2}\tau\right)[1+\mathcal{O}(1/J)]. \quad (5.15)$$

This means accelerated decoherence as long as $\gamma_1 \neq \gamma_2$ and $\gamma_1\gamma_2 \neq 1$. If, however, $\gamma_1\gamma_2 = 1$ then the next order in $1/J$ shows that

$$\frac{N_2(\tau)}{N_2(0)} = \exp\left(-\left(\frac{\gamma_1^2-1}{\gamma_1^2+1}\right)^2\tau - \frac{3\gamma_1^8 - 3\gamma_1^6 + 4\gamma_1^4 - 3\gamma_1^2 + 3}{2(\gamma_1^2+1)^4}\tau^2\right)$$
$$\times[1+\mathcal{O}(1/J)]. \quad (5.16)$$

The expression in the exponent is correct up to and including order $(J\tau)^2$. Obviously, accelerated decoherence is absent for $\gamma_1\gamma_2 = 1$. Indeed, a single coherent state $\gamma_1 = \gamma_2 = \gamma$ leads, in linear order, to almost the same decay,

$$\frac{N_2(\tau)}{N_2(0)} = \exp\left(-\gamma^4\left(\frac{\gamma^2-1}{\gamma^2+1}\right)^2\tau\right). \quad (5.17)$$

Figure 5.1 shows a comparison of (5.16) with numerical results obtained from the exact quantum mechanical propagator of the master equation (4.11). The agreement is very good, even up to times $\tau \sim 1$m, where one would not have expected (5.16) to be valid anymore. The reason is that, owing to the symmetry of the cat, the main contributions come from $m \simeq -n$ so that the precise condition of validity of the short-time propagator (4.23), $\tau \ll J/[|m+n-1|(n-m)]$, gives indeed $\tau \ll 1$, and not $\tau \ll 1/J$.

5.3.6 Preparation and Measurement

Let me dwell for a moment on the possibility of preparing the special Schrödinger cat states described above and of measuring their slow decoherence. It has been experimentally verified that (4.11) describes adequately the radiation by identical atoms resonantly coupled to a leaky resonator mode [108] in a suitable parameter regime (see Sect. 4.2.2). It should therefore in principle be possible to observe the slow decoherence of the special Schrödinger cat states. Let me first propose a scheme for their preparation.

Starting with all atoms in the ground state and with the field mode in its vacuum state, a resonant laser pulse brings the Bloch vector into a coherent state $|\theta,\phi\rangle$. Note that the cavity may be strongly detuned with respect to the atomic transition frequency during the whole preparation process (detuning $\delta \gg \kappa$). The dissipation mechanism (4.11) is turned off and by doing so the system evolves unitarily with a Hamiltonian containing a nonlinear term $\propto (g^2/\delta)J_+J_-$. The free evolution during a suitable time will split the coherent state $|\theta,\phi\rangle$ into a superposition of $|\theta,\phi'\rangle$ and $|\theta,\phi'+\pi\rangle$ (see [62]). Finally, a resonant $\pi/2$ pulse brings the superposition into the desired orientation symmetric with respect to the equator, by rotation through an angle $\pi/2$ about an axis perpendicular to the plane defined by the directions of the two coherent states produced by the free evolution. At this point the cavity

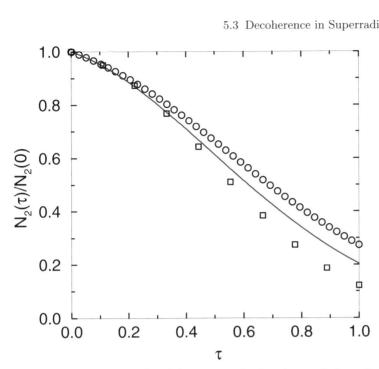

Fig. 5.1. Comparison of (5.16) (*continuous line*) with a result from direct numerical evaluation ($\gamma_1 = 0.5$, $\gamma_2 = 2.0$, $j = 10$ (*circles*) and $j = 20$ (*squares*)). The formula works well even up to times $\tau \simeq 1$

can be tuned to resonance, thereby switching on the dissipation mechanism, and one can study the decay of coherence. During all these manipulations spontaneous emission is, of course, not turned off and therefore presents a limiting timescale, compared with which all operations have to be performed quickly.

In order to study the decoherence process of the Schrödinger cat state it must also be detected. Unfortunately, the light emitted from the atoms in a Schrödinger cat state does not differ much from the light that is emitted from atoms in a single state. For example, if the state of the atoms is the polar cat state $(|j, -j\rangle + |j, j\rangle)/\sqrt{2}$, the intensity of the light is just reduced by a factor 2. So it is clear that one needs to measure the state of the atoms directly. In principle it is possible to measure density matrices for arbitrary systems by quantum tomography [147, 148, 149, 150, 151]. To what extent this is feasible in the current situation is an experimental question. The state of single atoms can be easily measured by selective ionization techniques [152] or selective fluorescence techniques [153].

5.3.7 General Decoherence-Free Subspaces

We have concentrated in this section so far on a small part of Hilbert space, namely the $(2j+1)$-dimensional subspace spanned by the states connected to initially totally excited atoms. We have seen that in this subspace there are pairs of almost decoherence-free states, namely angular-momentum coherent states that are arranged symmetrically with respect to the equator on a great circle. They are almost decoherence-free because they are (for $j \to \infty$) degenerate eigenstates of the coupling agent J_-, and only the system dynamics itself brings the state out of the decoherence-free subspace on a classical timescale. However, in the whole 2^N-dimensional Hilbert space, there is a much larger subspace that is *entirely* decoherence-free: it is the space of all states that fulfill

$$J_-|\psi\rangle = 0. \tag{5.18}$$

Freedom from decoherence arises because theses states are all degenerate eigenstates of J_- with the eigenvalue zero. The system dynamics itself does not change anything either, since the states are the ground states of all the representations with $j = 1/2, \ldots, N/2$. The dimension of this subspace is $\binom{N}{N/2}$ for even N and scales like $2^{N+1/2}/\sqrt{\pi N}$ for large N [134]. *The DFS is exponentially large in the number of atoms!* This result even holds for a more general coupling than (4.5), i.e. if the coupling constants g_i are different, $H_{\text{int}} = \sum_{i=1}^{N} \hbar g_i(\sigma_-^{(i)}b^\dagger + \sigma_+^{(i)}b)$, although the states in the DFS will then be different, of course.

Let me derive that surprising result. For brevity I shall assume a symmetric coupling $g_i = g$ for all atoms, but it will be clear in the end that the result generalizes for nonidentical g_i. A general state $|\psi\rangle$ of the atoms can be written as

$$|\psi\rangle = \sum_{\boldsymbol{\sigma}} c_{\boldsymbol{\sigma}} |\sigma_1 \ldots \sigma_N\rangle, \tag{5.19}$$

with arbitrary normalized coefficients $c_{\boldsymbol{\sigma}}$, and $\boldsymbol{\sigma} = (\sigma_1, \ldots, \sigma_N)$. The states $|\sigma_1 \ldots \sigma_N\rangle$ are product states $|\sigma_1 \ldots \sigma_N\rangle = |\sigma_1\rangle \ldots |\sigma_N\rangle$, where $\sigma_i = 0, 1$ labels the ground and excited states of atom i, and $\sigma_-^{(i)}|\sigma_i\rangle = \sigma_i|0\rangle$ for all atoms. Correspondingly, the state $|\psi'\rangle \equiv J_-|\psi\rangle$ can be expanded as

$$|\psi'\rangle = \sum_{\boldsymbol{\mu}} c'_{\boldsymbol{\mu}} |\mu_1 \ldots \mu_N\rangle. \tag{5.20}$$

In order to fulfill (5.18), the coefficients $c'_{\boldsymbol{\mu}} = \langle \boldsymbol{\mu} | \psi' \rangle$ must equal zero for all $\boldsymbol{\mu}$. One can verify in one line of calculation that this leads to the condition

$$c'_{\boldsymbol{\mu}} = \sum_{i=1}^{N} c_{\mu_1 \ldots \mu_{i-1} 1 \mu_{i+1} \ldots \mu_N} \delta_{\mu_i, 0} = 0 \tag{5.21}$$

for all $\boldsymbol{\mu}$. The number of different values of $\boldsymbol{\mu}$ indicates 2^N equations for the original coefficients $c_{\boldsymbol{\sigma}}$. Let us have a look at some of these equations.

1. Suppose that $\mu_l = 1$ for all l. Then (5.21) is automatically fulfilled, no restriction arises.
2. Suppose that one $\mu_l = 0$ and the rest equal one. Then (5.21) implies
$$c_{1\ldots111\ldots1} = 0, \tag{5.22}$$
where the 0 at position l has been replaced by a 1. So the DFS may not contain the highest excited state, as was to be expected. This condition arises N times.
3. If two μs are zero, say μ_l and μ_k (with $l \neq k$), (5.21) leads to
$$c_{1\ldots111\ldots101\ldots1} + c_{1\ldots101\ldots111\ldots1} = 0, \tag{5.23}$$
where in the first coefficient the 0 at position l has been replaced by a 1, and in the second coefficient the 0 at position k has been replaced by a 1. There are $\binom{N}{2}$ equations of this type, but they all couple only coefficients with exactly one σ_i equal to zero.
4. The general case with M different μ_ls equal to zero comprises $\binom{N}{M}$ equations that couple all coefficients with $M-1$ indices $\sigma_i = 0$. There are $\binom{N}{M-1}$ such coefficients for a given $\boldsymbol{\mu}$ and they appear only in the equations corresponding to this given M.

The total system of 2^N equations indexed by $\boldsymbol{\mu}$ decouples into blocks in which only the $c_{\boldsymbol{\sigma}}$s with $\boldsymbol{\sigma}$s that contain the same number of zeros are coupled. The binomial coefficient $\binom{N}{M}$ has a maximum at $M = N/2$ (I assume for simplicity that N is even; the case of N odd is treated just as easily). So, for $1 \leq M \leq N/2$, $\binom{N}{M} > \binom{N}{M-1}$, i.e. there are more equations than unknowns and there will be no solution besides the trivial one $c_{\boldsymbol{\sigma}} = 0$. So there are no decoherence-free states that contain more than half the maximum possible excitation.

If, on the other hand, $N/2 + 1 \leq M \leq N$, then $\binom{N}{M} < \binom{N}{M-1}$, i.e. there are more coefficients than equations, and $\binom{N}{M-1} - \binom{N}{M}$ coefficients remain undetermined and allow for decoherence-free states. So the dimension of the DFS is given by

$$\sum_{M=N/2+1}^{N} \left[\binom{N}{M-1} - \binom{N}{M}\right] = \sum_{k=1}^{N/2} \binom{N}{k} - \sum_{k=0}^{N/2-1} \binom{N}{k} \tag{5.24}$$

$$= \binom{N}{N/2} - \binom{N}{0}, \tag{5.25}$$

where, however, the ground state has not been counted yet. For the coefficient $c_{0\ldots0}$ never appears in the equations (5.21), since all coefficients contain at least one index equal to unity. But, as no conditions restrict $c_{0\ldots0}$, the ground state is of course always decoherence-free as well. Therefore, for N even, the dimension of the DFS is $d_{\text{DFS}} = \binom{N}{N/2}$, and for N odd one finds correspondingly $d_{\text{DFS}} = \binom{N}{(N+1)/2}$. Note that the result is indeed not restricted to $g_i = g$, since it is just based on counting equations and coefficients. If the

g_is are different, the linear combinations of coefficients change in each equation, but as long as no additional degeneracies are introduced, the dimension of the DFS will be the same. If additional equations become degenerate, the dimension can only increase.

5.4 Summary

Decoherence due to the coupling of a system to an environment is one of the main reasons why the macroscopic world that we live in appears to obey classical mechanics. Whereas the fundamental superposition principle of quantum mechanics allows the superposition of arbitrary states in Hilbert space, decoherence prevents, in general, observation of superpositions of macroscopically distinct states by destroying interference terms in extremely short times. While this general rule has been known for a long time, exceptions have only recently been discovered. I have presented in this chapter a detailed analysis of an exception in the case of superradiance that should in principle allow for an experimental observation. And I have pointed out possible applications in quantum computing by identifying a more general decoherence-free subspace, which, surprisingly, is exponentially large and contains roughly a fraction $\sqrt{2/\pi}/\sqrt{N}$ of the entire Hilbert space if the dimension 2^N of the latter is large enough.

6. Dissipative Quantum Maps

Dissipative quantum maps allow one to study the simultaneous effects of chaos, dissipation and decoherence in a relatively simple way. The interplay of these ingredients with quantum mechanics is very interesting and the subject of strong current research interest [42, 65, 66, 67, 68, 130, 131, 154, 155, 156, 157, 158]. We have seen that chaos manifests itself in rather different ways in quantum mechanics and classical mechanics. However, in the presence of dissipation and therefore decoherence, quantum mechanics becomes closer to classical mechanics, and classical aspects of chaos start to show up again. Phase space descriptions of quantum mechanics such as Wigner or Husimi functions therefore play a central role. All of this will be discussed in detail in the next chapter. The present short chapter serves to introduce dissipative quantum maps and to present the known properties of a dissipative kicked top, our model of choice for the rest of this book.

A noteworthy earlier study of dissipative quantum maps is the quantization of Henon's map by Graham and Tél [154]; Dittrich and Graham, as well as Miller and Sarkar, considered the kicked rotator with dissipation [68, 159]. References [9] and [160] treat numerically the transition from quantum mechanics to classical mechanics for the same kicked top with dissipation that I am going to present shortly. Miller and Sarkar have also studied an inverted harmonic oscillator with dissipation [158], as well as two coupled kicked rotators [67].

6.1 Definition and General Properties

A dissipative quantum map P is a map of a reduced density matrix ρ from a discrete time t to a time $t+1$,

$$\rho(t+1) = P\rho(t), \qquad (6.1)$$

where all eigenvalues of the propagator P that are not equal to unity have absolute values, smaller than unity.

The fact that P maps a density matrix to a density matrix immediately implies a series of consequences for P. Density matrices are Hermitian, have a trace equal to unity and are positive definite. The propagator has to conserve these properties. The consequences arising from this requirement have been largely exploited in [60, 161] and I shall therefore not derive but just state them here, with the exception of the implications of conservation of positivity, which seem to have gone unexplored so far.

The conservation of Hermiticity leads to an antiunitary symmetry of P, $[A, P] = 0$, where $AX = X^\dagger$ for all X. The existence of this symmetry implies that P can be given a real asymmetric representation. All eigenvalues are therefore real or come in complex conjugate pairs, and all traces of P are real.

Probability conservation implies that P has at least one eigenvalue equal to unity. The eigenstate corresponding to this eigenvalue is an invariant density matrix, and we shall see that the latter is very closely related to the invariant ergodic state of the Frobenius–Perron propagator $P_{\rm cl}$ of the corresponding classical map. In order to unravel the consequences of the conservation of positivity, let me write $\rho(t)$ and $\rho(t+1)$ in their respective eigenbases,

$$\rho(t) = \sum_{i=1}^{N} \rho_i(t) |u_i\rangle\langle u_i|,$$

$$\rho(t+1) = \sum_{l=1}^{N} \rho_l(t+1) |v_l\rangle\langle v_l|. \tag{6.2}$$

Positivity of $\rho(t)$ and $\rho(t+1)$ implies that the eigenvalues $\rho_i(t)$ and $\rho_l(t+1)$ are nonnegative. Furthermore, we have, for any density matrix, $\operatorname{tr}\rho(t) = 1 = \sum_{i=1}^{N} \rho_i(t)$. The eigenvalues therefore obey $0 \leq \rho_i(t) \leq 1$ for all times. Note, however, that the above representation is not the standard representation $\rho = \sum_i p_i |\psi_i\rangle\langle\psi_i|$, in which the p_i denote occupation probabilities of not necessarily orthogonal states ψ_i. Rather, we have made use of the Hermiticity of ρ, which guarantees an orthonormal set of eigenstates $|u_i\rangle$ and real eigenvalues ρ_i.

It is useful to introduce the Dirac notation $|u_{lm}\rangle\rangle = |u_l\rangle\langle u_m|$ since these are the basis states in operator space on which P acts [60]. To each ket there is a bra $\langle\langle u_{lm}| = (|u_l\rangle\langle u_m|)^\dagger = |u_m\rangle\langle u_l|$, and a scalar product can be defined as $\langle\langle u_{lm}|u_{kj}\rangle\rangle = \operatorname{tr}\left((|u_l\rangle\langle u_m|)^\dagger |u_k\rangle\langle u_j|\right) = \delta_{kl}\delta_{mj}$. One can easily check that it fulfills all required properties of a scalar product. If we introduce a similar notation for the final states, $|v_{lm}\rangle\rangle = |v_l\rangle\langle v_m|$, the action of P can be written as

$$\rho_j(t+1) = \sum_{i=1}^{N} \rho_i(t) \langle\langle v_{jj}|P|u_{ii}\rangle\rangle. \tag{6.3}$$

There are density matrices for which $\rho_i(t) = \delta_{ik}$ for some given k, i.e. they are just projectors onto the state $|u_k\rangle$. It follows that

$$\langle\langle v_{jj}|P|u_{kk}\rangle\rangle = \rho_j(t+1) \geq 0. \tag{6.4}$$

Thus, the propagator, when sandwiched between two arbitrary projectors, must always give a nonnegative value.

6.1.1 Type of Maps Considered

The type of maps that I consider in the following are particularly simple in the sense that the dissipation is well separated from a remaining purely unitary evolution, and the latter by itself is capable of chaos. The unitary part is described by a unitary Floquet matrix F as for an ordinary unitary quantum map (Chap. 3), and the dissipation by a propagator D, which necessarily acts on a density matrix. So the total map P is of the form

$$\rho(t+1) = DF\rho(t)F^\dagger \equiv P\rho(t). \tag{6.5}$$

Such a separation into two parts is not purely academic. A most obvious realization of (6.5) is obtained when the Hamiltonian $H(t)$ leading to the unitary evolution and the coupling to the environment can be turned on and off alternately (see below for a proposed experimental realization of the dissipative kicked top). Another example might be a billiard in which the particle only dissipates energy when hitting the walls – or the inverse situation, where the reflection from the walls is elastic, but the particle suffers from friction in the body of the billiard. But even if the dissipation cannot be turned off, the map (6.5) may still be a good description. For instance, if the dissipation is weak and if the entire unitary evolution takes place during a very short time, dissipation may be negligible during that time. This is the case if the entire unitary evolution is due to periodic kicking. The dissipation can then be considered as a relaxation process between two successive kicking events. Finally, a formal reason for such a separation can be given in the case where the generators of the unitary evolution and the dissipation commute. Technically, the separation of the map into a unitary evolution and a relaxation process simplifies greatly the analysis.

6.2 A Dissipative Kicked Top

In this section I introduce the primary example of a dissipative quantum map that will be studied throughout the rest of this book, and present an account of its known classical and quantum mechanical properties.

The dissipative kicked top of our choice is a map of the form (6.5), where F is the Floquet matrix (3.5) for the kicked top, and D the relaxation described by the master equation (4.11). The generator of the dissipation defined in (4.11) does not commute with $F(\cdot)F^\dagger$. In order to obtain the map (6.5) one must replace $H_0 = (k/2JT)J_z^2$ by $(k/2JT_1)J_z^2$ in (3.4) and switch on $H(t)$ only for a time $T_1 < T$ during each period T, whereas the dissipation acts during the rest of the time $\tau = T - T_1$.

The dissipative kicked top should be realizable, for instance, with atoms or ions flying through a series of cavities and laser beams which realize either the unitary evolution or the dissipation. Alternatively, one might think of keeping the atoms permanently in a cavity and use an AC Stark effect to tune them on and off resonance. In the following I shall assume the dissipative kicked top as given, and use (6.5) with F and D given by (3.5), (4.11) and (4.15) as a starting point for the subsequent theoretical analysis. The parameter τ will measure the relaxation time between two unitary evolutions and thus the dissipation strength.

6.2.1 Classical Behavior

To get a taste of the rich phenomenology of the dissipative kicked top, let me start by presenting an overview of its classical behavior as a function of dissipation strength. The classical map $(\mu, \phi) = \boldsymbol{f}_{\text{cl}}(\mu', \phi')$ corresponding to the dissipative part is defined by (4.10) with $\mu = \mu(\tau)$, $\mu' = \mu(0)$, $\phi = \phi(\tau)$ and $\phi' = \phi(0)$. This classical trajectory will be denoted by $\mu = \mu_{\text{d}}(\mu')$. Let me also give the explicit form of the unitary part. The rotation by an angle β about the y axis corresponds to [92]

$$\mu = \mu' \cos\beta - \sqrt{1 - (\mu')^2}\sin\beta \cos\phi', \tag{6.6}$$

$$\phi = \left\{ \arcsin\left(\sqrt{\frac{1-(\mu')^2}{1-\mu^2}}\sin\phi'\right)\theta(x) \right. \tag{6.7}$$

$$\left. + \left[\text{sign}(\phi')\pi - \arcsin\left(\sqrt{\frac{1-(\mu')^2}{1-\mu^2}}\sin\phi'\right)\theta(-x)\right] \right\} \bmod 2\pi,$$

$$x = \sqrt{1-(\mu')^2}\cos\phi'\cos\beta + \mu'\sin\beta, \tag{6.8}$$

where x is the x component of the angular momentum after rotation, $\theta(x)$ is the Heaviside theta function, $\text{sign}(x)$ denotes the sign function, and by arcsin the principal value in $]-\pi/2, \pi/2]$ is meant. To simplify the notation I have left the final μ in the right-hand side of (6.7) instead of replacing it via (6.6). For the torsion around the z axis we have simply

$$\mu = \mu', \tag{6.9}$$
$$\phi = (\phi' + k\mu') \bmod 2\pi. \tag{6.10}$$

Figure 6.1 shows phase space portraits of the classical dissipative kicked top with the parameters $k = 4.0$, $\beta = 2.0$ for several values of τ. Without dissipation ($\tau = 0$), the phase space portrait is almost structureless, in agreement with the earlier statement that simultaneously large values of k and β lead to chaotic behavior. Two tiny but visible stability islands remind us of the fact that at $k = 0$ the behavior was integrable. Upon increasing the dissipation strength, the fixed points of the map become repellers or attractors, and even at small values of τ, typically of the order $\tau \simeq 0.1$, a strange attractor arises [162]. The attractor shrinks more and more and at the same time sinks towards the south pole $\mu = -1$ when the dissipation is increased. The dimension reduces finally to zero (see Fig. 6.2), and the strange attractor becomes a point attractor, typically accompanied by a point repeller further up on the

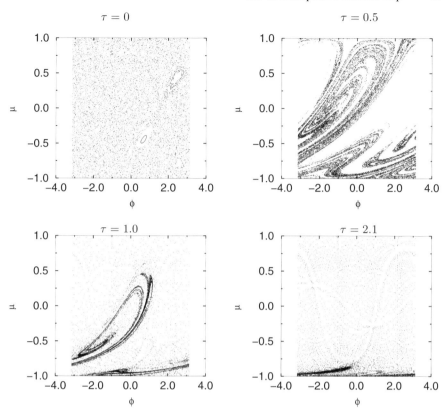

Fig. 6.1. Strange attractors for a dissipative kicked top with $k = 4.0$, $\beta = 2.0$ and various values of τ

sphere. The latter is not shown in the figure, but can be made visible by inverting the map. The stage of a point attractor is typically reached at $\tau \gtrsim 3$. Dissipation is so strong here that the system has become integrable again: all trajectories jump immediately into the point attractor. For intermediate values of τ, fixed points and the strange attractor coexist peacefully.

The evolution of the Lyapunov exponent as a function of τ can be seen in Fig. 6.3. For zero dissipation and $\beta \neq 0$, the system is chaotic if $k > 0$, and is more chaotic for larger k. With increasing τ, the Lyapunov exponent typically decreases, although the behavior is not entirely monotonic. For some finite value of τ depending on the torsion strength k, the system becomes integrable. For $k = 4.0$ there is a small region of reentrant chaotic behavior, though. Since the dissipation leads to a nonlinear classical evolution as well, one might wonder whether dissipation and rotation alone cannot lead to chaotic behavior. That this is not the case is also shown in the figure: for $k = 0$ the Lyapunov exponent remains zero, till at $\sinh(\tau/2) = \tan(\beta/2)$

68   6. Dissipative Quantum Maps

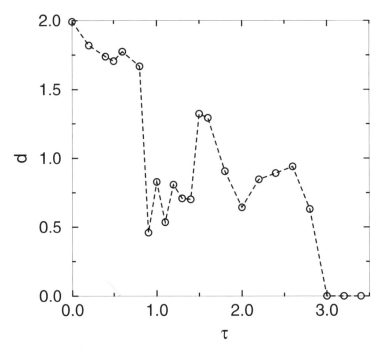

Fig. 6.2. Box-counting dimension of the strange attractor for a dissipative kicked top with $k = 4.0$, $\beta = 2.0$ as a function of dissipation. For $\tau = 0$ the two-dimensional phase space is uniformly filled; for large dissipation a point (attractor) of zero dimension occurs. In between, the dimension of the strange attractor decays in a rather erratic way

a bifurcation is reached and a point attractor/repeller pair is born [163], whereupon the Lyapunov exponent becomes negative.

6.2.2 Quantum Mechanical Behavior

Quantum mechanically, the dissipative quantum map is characterized by the eigenvalues and eigenstates of P. They allow for a clear distinction between several damping regimes. Let me focus here on the eigenvalues, which are more easily accessible.

- At zero dissipation, $\tau = 0$, the dissipative propagator becomes the unity operator, i.e. the quantum map (6.5) is a unitary quantum map formulated for density matrices, $\rho(t+1) = F\rho(t)F^\dagger$.
  Suppose that F has eigenvectors $|u_l\rangle$, $l = -j, -j+1, \ldots, j$. The corresponding eigenvalues λ_l are unimodular, $\lambda_l = e^{i\varphi_l}$, since F is unitary. It immediately follows that $P(\tau = 0)$ has eigenstates $|u_l\rangle\langle u_k|$ with eigenvalues

$$\lambda_{l,k} = e^{i(\varphi_l - \varphi_k)}, \quad l, k = -j, \ldots, j. \tag{6.11}$$

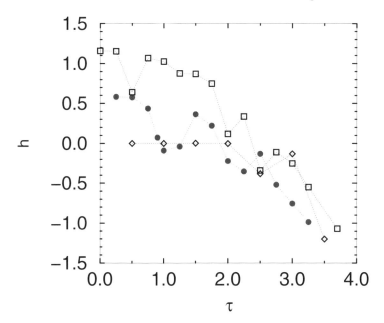

Fig. 6.3. Lyapunov exponent for a dissipative kicked top with $\beta = 2.0$ as a function of dissipation for different torsion strengths ($k = 8.0$, *squares*; $k = 4.0$, *filled circles*; $k = 0$, *diamonds*). The Lyapunov exponent decreases overall with dissipation strength, and the system becomes integrable for $h \leq 0$. For $k = 4.0$ there is a small region of reentrant chaotic behavior. For $k = 0$, h remains zero, till at $\tau \simeq 2$ a bifurcation is reached in which a point attractor/repeller pair is born. The *dotted lines* are guides to the eye only

Any two eigenvalues $\lambda_{l,k}$ and $\lambda_{k,l}$ are complex conjugate, and there are $2j + 1$ degenerate eigenvalues $\lambda_{l,l} = 1$. Since all eigenvalues lie on the unit circle, the mean level spacing between the eigenphases $(\varphi_l - \varphi_k)$ is $2\pi/(2j+1)^2$, i.e. of the order of $1/J^2$. Out of the $(2j+1)^2$ eigenphases, level repulsion is only present within groups consisting of $(2j+1)$ members if the system is chaotic, namely between all eigenvalues with fixed l (and k running from $-j$ to j). There is no need for eigenphases $(\varphi_l - \varphi_k)$ that have different l *and* k to repel. The statistics of the eigenphases at zero dissipation are therefore the Poisson statistics of uncorrelated numbers on a line, with a nearest-neighbor distance distribution $P(s) = \exp(-s)$.

- For $\tau \ll 1/J^3$, the effect of the dissipation can be described perturbatively, since $D = \exp(\Lambda\tau) \simeq \mathbf{1} + \Lambda\tau + \ldots$ is very close to unity. The change of the eigenvalues of P is, to first order in τ, given by the expectation value of the perturbation Λ in the unperturbed eigenstates. Using the notation introduced in Sect. 6.1, the expectation value of Λ is given by $\Lambda_{lk} = \langle\langle U_{lk}|\Lambda|U_{lk}\rangle\rangle$, and consequently

$$\lambda_{l,k} \simeq e^{i(\varphi_l - \varphi_k)}(1 + \Lambda_{lk})$$

$$\simeq e^{i(\varphi_l - \varphi_k) + \Lambda_{lk}}. \tag{6.12}$$

It can be shown [162] that the average matrix element Λ_{lk} is roughly

$$\Lambda_{lk} \simeq -\frac{2}{3} J\tau. \tag{6.13}$$

The perturbation theory works as long as the eigenvalues do not move a distance comparable to or larger than a mean level spacing, hence the restriction to $\tau \ll 1/J^3$. In this regime the area in which the eigenvalues live is still almost one-dimensional.

- For $1/J^3 \sim \tau$, the eigenvalues have moved a distance comparable to the mean unperturbed level spacing into the inside of the circle. Their density has now two-dimensional support, but since dissipation is weak, dissipation does not introduce appreciable correlations yet. The statistics of the eigenvalues are therefore those of uncorrelated random numbers in the plane. It is well known that this leads exactly to the Wigner surmise $P(s) = (\pi/2) s \exp[-(\pi/4) s^2]$ for the statistics of the next-nearest-neighbor distance (which is now defined as a distance in the complex plane!) [162].
- When the dissipation has become so strong that the eigenvalues have moved to the inside of the circle, $J\tau \sim 1$, the eigenvalue statistics are drastically changed. In this regime, the so-called Ginibre statistics are observed, if the system is chaotic. These are the statistics of random non-Hermitian matrices with independently Gaussian-distributed entries [164] (see next section). This leads to cubic level repulsion, i.e. $P(s) \propto s^3$ for small s. The level repulsion is insensitive to the fundamental symmetries of the system such as (generalized) time reversal invariance or invariance under spin rotation. For integrable classical dynamics the nearest-neighbor statistics are still linear for small s.

Not much is known about the transition regime between $\tau \sim 1/J^3$ and $\tau \sim 1/J$. A transition from uncorrelated to correlated spectra has to take place. Note that the Ginibre statistics depend on a quantum mechanical scale for the dissipation, $\tau \gtrsim 1/J$. Therefore, a chaotic system that shows classically a strange attractor may have different eigenvalue statistics depending on the dimension of the Hilbert space (and therefore the value of the effective $\hbar$) chosen. For instance, at $\tau = 0.05$, which is strong enough to give rise to a weak strange attractor, a quantum mechanical system with $j = 2$ has $\tau < 1/J^3$ and will therefore have uncorrelated eigenvalues almost on the unit circle. (Of course, the 25 eigenvalues are not enough for the calculation of $P(s)$, but one may vary k a little, for example, and study level correlations by the motion of the levels). But for $j = 50$, clearly $\tau > 1/J$, and we therefore obtain Ginibre statistics for the spacings. In general, for any arbitrarily small but finite τ we expect Ginibre statistics for large enough values of j, if the system is chaotic.

We conclude that Ginibre statistics set in not with classical dissipation but with quantum mechanical decoherence. As we have seen in Chap. 5, decoherence *does* typically happen on a scale $\tau \sim 1/J$. On the other hand, dissipation should not be too strong either, since for large values of τ, $\tau \gtrsim 1$, the system becomes integrable again and we should not expect correlated eigenvalues.

In the following chapter I shall analyze the regime $\tau > 1/J$ semiclassically. We shall make use of the propagators developed in Chap. 4. Before doing so, let me say a few more words about Ginibre's ensemble.

6.3 Ginibre's Ensemble

In 1965 Ginibre introduced an ensemble of general complex matrices $\mathbf{S}$ [164]. He also considered general real and general quaternion matrices, but the results are more cumbersome in these cases [165]. Ensembles of non-Hermitian matrices have recently found renewed interest, as applications in different fields of physics were found [166, 167, 168, 169, 170]. As an ensemble of complex matrices describes very well the correlations of eigenvalues of P, I shall restrict myself to complex matrices. It has in fact been shown that cubic level repulsion is independent of whether the matrices $\mathbf{S}$ are asymmetric real, symmetric complex or general complex [161].

The measure in the matrix space can be defined by

$$\mathrm{d}\mu(\mathbf{S}) = \prod_{i,j} \mathrm{d}\mathrm{Re}\, S_{ij} \mathrm{d}\mathrm{Im}\, S_{ij} \mathrm{e}^{-\mathrm{tr}\, \mathbf{SS}^\dagger}, \qquad (6.14)$$

where $\mathrm{Re}\, S_{ij}$ and $\mathrm{Im}\, S_{ij}$ are the real and imaginary parts, respectively, of the matrix element S_{ij}, and i,j run from 1 to N, if N is the dimension of the matrix. The measure is invariant under unitary rotations and leads to statistically independent matrix elements. Ginibre calculated the N-point joint probability distribution function $P_N(z_1, \ldots, z_N)$ of the complex eigenvalues $z_1, \ldots, z_N$, as well as reduced joint probability densities which arise from integrating out some of the variables. Deriving these results here would be beyond the scope of the present book; so let me just quote them. The interested reader is invited to check out the original reference [164].

The joint probability distribution function reads

$$P_N(z_1, \ldots, z_N) = \mathcal{N}^{-1} \prod_{i<j} |z_i - z_j|^2 \mathrm{e}^{-\sum_i |z|_i^2}, \qquad (6.15)$$

where the normalization constant $\mathcal{N}$ is given by

$$\mathcal{N} = \prod_{k=1}^{N} (\pi k!). \qquad (6.16)$$

This constant assures that $\int \prod_{i=1}^{N} \mathrm{d}^2 z_i\, P_N(z_1, \ldots, z_N) = 1$, where, for the decomposition $z = x + iy$ into real and imaginary parts, $\mathrm{d}^2 z \equiv \mathrm{d}x\, \mathrm{d}y$. By integrating out some of the variables in (6.15), the reduced joint probability densities (joint densities for short)

$$\rho(z_1,\ldots,z_n) \equiv \int P_N(z_1,\ldots,z_N)\,d^2z_{n+1}\ldots d^2z_N \tag{6.17}$$

can be derived. With the help of the "incomplete exponential function"

$$e_N(x) \equiv \sum_{l=0}^{N} \frac{x^l}{l!}, \tag{6.18}$$

the result takes the simple form [60]

$$\rho(z_1,\ldots,z_n) = \frac{(N-n)!}{N!} \frac{e^{-\sum_i |z_i|^2}}{\pi^n} \det\left[e_{N-1}(z_i z_k^*)\right], \tag{6.19}$$

with $i,k = 1,\ldots,n$. In particular, the mean eigenvalue density $\rho(z)$ in the complex plane is given by

$$\rho(z) = \frac{1}{\pi N} e^{-|z|^2} e_{N-1}(|z|^2). \tag{6.20}$$

The combination of the exponential and "incomplete exponential" functions behaves for large N like an error function,

$$e^{-x} e_N(x) \simeq \frac{1}{2}\text{erfc}\left(\frac{x-N}{\sqrt{2N}}\right), \tag{6.21}$$

and produces a sharp edge at $x \simeq N$, in the sense that the function vanishes on a scale $\sqrt{2N}$ for any x larger than N. For $N \to \infty$, the density of states therefore becomes uniform in a circle of radius $|z| = \sqrt{N}$;

$$\rho(z) = \frac{1}{\pi N} \begin{cases} 1 & \text{for } |z| \leq \sqrt{N} \\ 0 & \text{otherwise} \end{cases}. \tag{6.22}$$

By rescaling z according to $z = \zeta\sqrt{N}$, one obtains a well-defined limiting density for the ζs, which is $1/\pi$ within the unit circle and zero elsewhere.

The two-point correlation function $\rho(z_1, z_2)$ can be read off from (6.17) just as easily. It vanishes in the limit $N \to \infty$ whenever any of the arguments has an absolute value larger than $\sqrt{N}$, and is otherwise given by

$$\rho(z_1, z_2) = \frac{1}{\pi^2 N(N-1)} \left(1 - e^{-|z_1-z_2|^2}\right). \tag{6.23}$$

We observe a "correlation hole" for small distances $|z_1 - z_2| \ll 1$, where $\rho(z_1, z_2)$ vanishes quadratically, i.e. $\rho(z_1, z_2) \propto |z_1 - z_2|^2$. This is of course a remnant of the quadratic term in (6.15). As in the case of real eigenvalues, the two-point correlation function is dominated at small values of $s \equiv |z_1 - z_2|$ by the nearest-neighbor-spacing distribution function $P(s)$. However, since s is now a distance in the complex plane, $P(s)$ is not just given by $R(s) \equiv \rho(z_1, z_2)\,|_{|z_1-z_2|=s}$, but acquires an additional factor s from the two-dimensional volume element $s\,ds\,d\varphi$, so that for small values of s, $P(s) \propto s^3$.

This is the cubic level repulsion for the Ginibre ensemble mentioned earlier. Recently the eigenvector statistics resulting from Ginibre's ensemble have also been calculated [171, 172], but there has been no comparison with the statistics for the dissipative kicked top yet.

6.4 Summary

I have defined in this section what I mean by a dissipative quantum map. As example that will be analyzed semiclassically in the next chapter, a dissipative kicked top was introduced, and I have given an overview of its known quantum mechanical and classical behavior. A small amount of dissipation τ leads, classically, always to a strange attractor, whereas the quantum mechanical behavior depends on the value of the quantum number j. For large enough j ($\tau \gtrsim 1/j$) one always has Ginibre statistics, whereas $\tau \sim 1/j^3$ gives Poisson statistics in the plane, and $\tau \ll 1/j^3$ gives Poisson statistics on the unit circle. For classically strong dissipation, the system becomes integrable again, as all trajectories run very rapidly into strongly attracting regions in phase space.

7. Semiclassical Analysis of Dissipative Quantum Maps

In this chapter I show how a great variety of information on dissipative quantum maps of the type (6.5) introduced in the previous chapter can be gained with semiclassical methods. I shall focus on

- the spectrum of the quantum propagator P
- the invariant state of the quantum map, i.e. the density matrix ρ that fulfills $P\rho = \rho$
- the time evolution of quantum mechanical expectation values
- correlation functions of observables.

The semiclassical methods that will be used are based heavily on those introduced in Sects. 3.4.1 and 4.4. There we have seen how the propagators for unitary quantum maps and for a purely dissipative relaxation process can be obtained semiclassically. What remains to be done is to combine these propagators to obtain the total propagator P and to extract the desired information. As technical tools, Poisson summation and saddle-point integration will be used. Owing to the combined effects of unitary motion (with an imaginary exponent in the propagator) and relaxation (with a purely real exponent), we shall encounter saddle-point integrals with a complex exponent depending on many variables. The correct treatment of such integrals, and in particular the determining of the overall phase factor, presents considerable technical difficulty and the corresponding mathematics is not easily found in the literature. Appendix A is therefore devoted to this mathematical problem, in an attempt to make the presentation self-contained.

7.1 Semiclassical Approximation for the Total Propagator

Let us start by deriving a semiclassical approximation for the total propagator P defined by (6.5). I first write down an exact formal expression in terms of matrix elements of the Floquet matrices F and $F^\dagger$, and of the propagator for the relaxation process D. The latter was defined by (4.11) and (4.15) in Chap. 4, the former by (3.5) in Chap. 3.

Immediately after the unitary motion induced by F and $F^\dagger$, ρ has the matrix elements

76   7. Semiclassical Analysis of Dissipative Quantum Maps

$$\rho_n(k,0+) = \langle n_1|F\rho(0)F^\dagger|n_2\rangle = \sum_{l_1,l_2=-j}^{j} F_{n_1 l_1}(F^\dagger)_{l_2 n_2}\langle l_1|\rho(0)|l_2\rangle, \quad (7.1)$$

where n_1, n_2 and l_1, l_2 are still the ordinary J_z quantum numbers ranging from $-j$ to j. The matrix elements of $\rho_n(k,0+)$ are labeled as in Chap. 4 by the "center of mass" index $n = (n_1 + n_2)/2$ and half the relative index $k = (n_1 - n_2)/2$. Introducing similarly $m' = (l_1 + l_2)/2$ and $k' = (l_1 - l_2)/2$ we can rewrite the sums, and find

$$\rho_n(k,0+) = \sum_{m',k'} F_{n+k,m'+k'} F^*_{n-k,m'-k'} \rho_{m'}(k',0). \quad (7.2)$$

This density matrix serves as a starting point for the relaxation process with the propagator D. So we insert $\rho_n(k,0+)$ into the right-hand side of (7.2), and obtain

$$\rho_m(k,\tau) = \sum_{n,m',k'} D_{mn}(k,\tau) F_{n+k,m'+k'} F^*_{n-k,m'-k'} \rho_{m'}(k',0)$$

$$\equiv \sum_{m',k'} P_{mk;m'k'} \rho_{m'}(k',0). \quad (7.3)$$

We read off the propagator P,

$$P_{mk;m'k'} = \sum_n D_{mn}(k,\tau) F_{n+k,m'+k'} F^*_{n-k,m'-k'}, \quad (7.4)$$

so far an exact representation in terms of the propagators F and D written in "center of mass" and "half relative" indices.

Now the first semiclassical approximation comes in. We replace F, $F^\dagger$ and D by the semiclassical expressions (3.11) and (4.64). The sum over n in (7.4) can be performed by Poisson summation,

$$\sum_{n=-\infty}^{\infty} f_n = \sum_m \int dn\, f(n) e^{i2\pi mn}, \quad (7.5)$$

and subsequent integration via the saddle-point approximation (SPA). Before the integration is done, an arbitrary matrix element of P reads

$$P_{mk;m'k'} = \sum_{l=-\infty}^{\infty} \int d\nu \sum_{\sigma_1,\sigma_2} B(\mu,\nu;\eta) C_{\sigma_1}(\nu+\eta,\mu'+\eta')$$
$$\times C^*_{\sigma_2}(\nu-\eta,\mu'-\eta') \exp\left[JG(\mu,\eta;\mu',\eta';\nu)\right], \quad (7.6)$$

where again $m = \mu J$, $n = \nu J$, $k = \eta J$, etc., and l is the integer in the Poisson summation. The indices σ_1 and σ_2 label the paths in the unitary step. The prefactor $C(\mu,\nu)$ is defined by (3.11) as

$$C_\sigma(\nu,\mu) = \frac{1}{\sqrt{2\pi J}} \sqrt{|\partial_\nu \partial_\mu S_\sigma(\nu,\mu)|}, \quad (7.7)$$

7.1 Semiclassical Approximation for the Total Propagator

and $B(\mu, \nu; \eta)$ is given by (4.67). I have dropped the dependence on τ, since τ is now just a fixed system parameter that measures the dissipation strength. Similarly, the dependence on the parameters k and β for the unitary motion will not be displayed explicitly.

The total "action" G is composed of a real part R from the relaxation process, and two imaginary parts from F and $F^\dagger$,

$$G(\mu, \eta; \mu', \eta'; \nu) \qquad (7.8)$$
$$= R(\mu, \nu; \eta) + iS_{\sigma_1}(\nu + \eta, \mu' + \eta') - iS_{\sigma_2}(\nu - \eta, \mu' - \eta') + i2\pi l\nu.$$

In order to evaluate the integral over ν by the SPA we have to solve the saddle-point equation

$$\partial_\nu G = 0 \qquad (7.9)$$
$$= \partial_\nu R(\mu, \nu; \eta) + i\left[\partial_\nu S_{\sigma_1}(\nu + \eta, \mu' + \eta') - \partial_\nu S_{\sigma_2}(\nu - \eta, \mu' - \eta') + 2\pi l\right].$$

In general, this equation will have complex solutions $\nu = \bar{\nu}(\mu, \eta; \mu', \eta')$. It may even be possible that several solutions or no solution at all exists. However, note that (7.9) simplifies considerably and gives a physically meaningful saddle point if $\sigma_1 = \sigma_2$ and $\eta = \eta' = l = 0$ (and that is the situation we shall encounter soon): we then have $\partial_\nu R(\mu, \nu; 0) = 0$, which is equivalent to the classical dissipative equation of motion $\bar{\nu} = \mu_d^{-1}(\mu)$ at an energy of the fictitious underlying Hamiltonian system $\tilde{E} = 0$ (see Chap. 4).

When the ν integral is evaluated by the SPA, the second derivative $\partial_\nu^2 G|_{\nu=\bar{\nu}} \equiv \partial_{\bar{\nu}}^2 G$ appears in the prefactor; it can be combined with the other preexponential factors in P. One needs a relation between second derivatives of G, which can be obtained by differentiating with respect to μ the saddle-point equation (7.9), accounting for the μ dependence of $\bar{\nu}$;

$$\partial_{\bar{\nu}}^2 G = -\partial_\mu \partial_{\bar{\nu}} G \frac{1}{\partial \bar{\nu}/\partial \mu} = -\left(\frac{\partial \bar{\nu}}{\partial \mu}\right)^{-1} \partial_\mu \partial_{\bar{\nu}} R. \qquad (7.10)$$

With the abbreviation $\psi(\mu, \eta; \mu', \eta') = G[\mu, \eta; \mu', \eta'; \bar{\nu}(\mu, \eta; \mu', \eta')]$ for the action at the saddle point and with (4.67), we find

$$P_{mk;m'k'} = \sum_{l=-\infty}^{\infty} \sum_{\sigma_1,\sigma_2} \sum_{\bar{\nu}} \sqrt{\left(\frac{\partial \bar{\nu}}{\partial \mu}\right)_{\tilde{E}} \left(\frac{\partial \bar{\nu}}{\partial \mu}\right)} \qquad (7.11)$$
$$\times C_{\sigma_1}(\bar{\nu} + \eta, \mu' + \eta') C^*_{\sigma_2}(\bar{\nu} - \eta, \mu' - \eta') \exp\left[J\psi(\mu, \eta; \mu', \eta')\right].$$

The sum over $\bar{\nu}$ picks up all relevant saddles. Note that of the two factors under the second square root, only the first one is at constant $\tilde{E}$. However, for $\eta = \eta' = 0$ and classical trajectories ($\tilde{E} = 0$), the two become identical and combine to give the classical Jacobian. That is precisely the situation that we are going to encounter soon.

With (7.11), we have a semiclassical approximation for the total propagator P in our hands that allows for the analytical calculation of many quantities that until now have been accessible to numerical evaluation only.

7.2 Spectral Properties

7.2.1 The Trace Formula

We have seen in Chap. 3 that the knowledge of N traces of any N-dimensional matrix suffices, at least in principle, for calculating all the eigenvalues of the matrix, since the traces determine uniquely the characteristic polynomial. As perhaps the simplest spectral properties, I therefore propose now to calculate traces of arbitrary integer powers P^t of the propagator P. This was done for the first time in [163]. Even though the result is simple, the calculation turns out to be quite cumbersome. I shall therefore not repeat the calculation in all details here, but rather show how one proceeds in principle. Also, I shall present directly the calculation of $\operatorname{tr} P^t$ for $t \geq 2$. The cases $t = 1, 2$ must be treated separately, as I shall use a determinant relation valid only for $t \geq 3$, but the final result will also be valid for $t = 1, 2$. Readers interested in the details of the calculation or in the cases $t = 1, 2$ are invited to consult the original paper [163]. A much simpler, though slightly less rigorous method that leads to the same result will be accessible after I will have introduced the Wigner propagator in Sect. 7.3.

The starting point for the calculation of $\operatorname{tr} P^t$ is the exact representation

$$\operatorname{tr} P^t = \sum_{m_1,k_1,\ldots,m_t,k_t} P_{m_1 k_1; m_t k_t} P_{m_t k_t; m_{t-1} k_{t-1}} \cdots P_{m_2 k_2; m_1 k_1} . \tag{7.12}$$

The sums are transformed into integrals by Poisson summation; the corresponding integers will be called r_i and t_i ($i = 1, \ldots, t$). To avoid problems arising from the fact that m and k can be simultaneously half-integers, it is useful to use $m' \equiv m + k$ and $2k$ as summation variables. Transforming back to m and k in the integral leads then to a prefactor 2 for each pair m_i, k_i. I insert in (7.12) the semiclassical approximation (7.11) for the propagators P, such that $m_i = J\mu_i$, $k_i = J\eta_i$, and $\bar{\nu}_i$ are the intermediate coordinates in the ith step of the map. It is convenient to introduce periodic boundary conditions for all variables, i.e. $\mu_{i+t} = \mu_i$, etc. These conditions are implied by the trace operation, and taking them into account avoids special treatment of the variables with index $t+1$. I shall also introduce a vector notation $\boldsymbol{\mu} = (\mu_1, \ldots, \mu_t)$, $\boldsymbol{\eta} = (\eta_1, \ldots, \eta_t)$, etc. In the case of the path indices σ, the first index will count the iteration of the map and the second will indicate whether it refers to a path in F or $F^\dagger$, $\boldsymbol{\sigma}_1 = (\sigma_{11}, \sigma_{21}, \ldots, \sigma_{t1})$ and $\boldsymbol{\sigma}_2 = (\sigma_{12}, \sigma_{22}, \ldots, \sigma_{t2})$. The integral representation of $\operatorname{tr} P^t$ has the form

$$\operatorname{tr} P^t = (2J^2)^t \int d\mu_1 \, d\eta_1 \ldots d\mu_t \, d\eta_t \sum_{r,l,t} \exp\left[J\Psi_t(\boldsymbol{\mu}, \boldsymbol{\eta})\right]$$

$$\times \prod_{i=1}^{t} \left[\sqrt{\left(\frac{\partial \bar{\nu}_i}{\partial \mu_{i+1}}\right)_{\tilde{E}} \left(\frac{\partial \bar{\nu}_i}{\partial \mu_{i+1}}\right)} C_{\sigma_{i1}}(\bar{\nu}_i + \eta_{i+1}, \mu_i + \eta_i) \right.$$

$$\left. \times C^*_{\sigma_{i2}}(\bar{\nu}_i - \eta_{i+1}, \mu_i - \eta_i) \right],$$

7.2 Spectral Properties

where the total "action" is given by

$$\Psi_t(\boldsymbol{\mu}, \boldsymbol{\eta}) = \sum_{i=1}^{t} \Big(\psi(\mu_{i+1}, \eta_{i+1}; \mu_i, \eta_i) + i2\pi[r_i(\mu_i + \eta_i) + 2t_i\eta_i]\Big)$$

$$= \sum_{i} \Big(R_i + i\big(S_{\sigma_{i1}} - S_{\sigma_{i2}} + 2\pi[r_i(\mu_i + \eta_i) + 2t_i\eta_i]\big)\Big). \quad (7.13)$$

I have used the abbreviations $R_i = R(\mu_i, \bar{\nu}_{i-1}, \eta_i)$, $S_{\sigma_{i1}} = S_\sigma(\bar{\nu}_i + \eta_{i+1}, \mu_i + \eta_i)$ and $S_{\sigma_{i2}} = S_\sigma(\bar{\nu}_i - \eta_{i+1}, \mu_i - \eta_i)$. In the case of $S_{\sigma_{i1}}$ and $S_{\sigma_{i2}}$ the index i serves two purposes, namely as the index of σ but also to indicate the arguments of S_σ. It will therefore be kept, even if we find soon $\sigma_{i1} = \sigma_{i2} = \sigma$. We now integrate by the saddle-point approximation, using the fact that $J \to \infty$ in the semiclassical limit. The saddle-point equations read, for $k = 1, \ldots, t$,

$$\partial_{\mu_k}\Psi_t = \partial_{\mu_k}R_k + i\big(\partial_{\mu_k}S_{\sigma_{k,1}} - \partial_{\mu_k}S_{\sigma_{k,2}} + 2\pi r_k\big) = 0, \quad (7.14)$$

$$\partial_{\eta_k}\Psi_t = \partial_{\eta_k}R_k + i\Big(\partial_{\bar{\nu}_{k-1}}S_{\sigma_{k-1,1}} + \partial_{\bar{\nu}_{k-1}}S_{\sigma_{k-1,2}}$$

$$+ \partial_{\mu_k}S_{\sigma_{k,1}} + \partial_{\mu_k}S_{\sigma_{k,2}} + 2\pi(r_k + 2t_k)\Big) = 0. \quad (7.15)$$

Note that there are no terms arising from the dependence of $\bar{\nu}$ on μ_k and η_k. The reason is that, by construction, $\partial_{\bar{\nu}_l}\Psi_t = 0$ for all l.

Let us look for *real* solutions. Complexity would indeed look unphysical, as $m = \mu J \in \mathcal{Z}$, etc. Formally, I cannot exclude the existence of complex solutions. But we shall see that demanding real solutions leads to classical trajectories. And as long as classical solutions exist we expect them to dominate over nonclassical ones, since they do so even in nondissipative quantum mechanics, and dissipation should favor classical solutions even more.

The real and imaginary parts of (7.14) and (7.15) must then separately equal zero. From the real parts we learn that $\partial_{\eta_k}R = 0$ and therefore $\eta_k = 0$, and $\partial_{\mu_k}R = 0$. The latter equation means that μ_k and $\bar{\nu}_{k-1}$ are connected by a classical trajectory for the dissipative part of the map, $\mu_k = \mu_d(\bar{\nu}_{k-1})$, and that the azimuths before and after the kth dissipative step agree, $\phi_k = \phi'_k$.

The imaginary part leads to a meaningful message if we remember the generating properties of the actions $S_{\sigma_{k1}}$ and $S_{\sigma_{k2}}$. These properties can now be easily employed, since $\eta_k = 0$, so that the arguments of $S_{\sigma_{k1}}$ are just $(\bar{\nu}_k, \mu_k)$. The imaginary part of (7.14) gives

$$\phi^i_{\sigma_{k1}}(\bar{\nu}_k, \mu_k) - \phi^i_{\sigma_{k2}}(\bar{\nu}_k, \mu_k) + 2\pi r_k = 0. \quad (7.16)$$

So the initial azimuthal angles $\phi^i_{\sigma_{k1}}$ and $\phi^i_{\sigma_{k2}}$ agree (modulo 2π). Furthermore, the initial and final momenta $\bar{\nu}_k$ and μ_k also agree. But since one and only one trajectory originates from a given phase space point $(\phi^i_{\sigma_{k1}}, \mu_k) = (\phi^i_{\sigma_{k2}}, \mu_k)$, the two trajectory fragments σ_{k1} and σ_{k2} must be the same, i.e. $\sigma_{k1} = \sigma_{k2} \equiv \sigma_k$. Counting all angles in the interval $-\pi \ldots \pi$ implies $r_k = 0$. From the imaginary part of (7.15) we obtain

$$-2\phi^f_{\sigma_{k-1}} + 2\phi^i_{\sigma_k} + 4\pi t_k = 0, \quad (7.17)$$

i.e. $\phi^{\text{f}}_{\sigma_{k-1}}$ and $\phi^{\text{i}}_{\sigma_k}$ have to agree modulo 2π. So we have found that in each step of the map, only classical trajectories that are implied by the generating properties of the actions contribute. The segments of these trajectories form a closed loop in phase space, i.e. a periodic orbit. Indeed, the coordinates transform in a sequence of t unitary and dissipative steps, U_i and D_i, according to

$$(\mu_1, \phi^{\text{i}}_{\sigma_1}) \xrightarrow{U_1} (\bar{\nu}_1, \phi^{\text{f}}_1) \xrightarrow{D_1} (\mu_2, \phi^{\text{f}}_1 = \phi^{\text{i}}_{\sigma_2}) \xrightarrow{U_2} \ldots . \tag{7.18}$$

Since exactly one trajectory originates from a given phase space point $(\mu_1, \phi^{\text{i}}_{\sigma_1})$, we can relabel all trajectory segments by the same label $\sigma_k \equiv \sigma$. To simplify the notation, I shall drop the index σ in the following altogether, but the reader should bear in mind that all actions will have to be evaluated at the periodic point σ. The index k will be kept in order to indicate the arguments of S, i.e. $S_k = S(\bar{\nu}_k, \mu_k)$.

To proceed further with the saddle-point approximation, we need the $2t \times 2t$ matrix $\mathbf{Q}_{2t \times 2t}$ of second derivatives of Ψ_t. It consists of blocks (μ, μ), (μ, η), (η, μ) and (η, η), where (μ, η) contains the mixed derivatives $-\partial_{\mu_i} \partial_{\eta_j} \Psi_t$ (and correspondingly for the other blocks):

$$\mathbf{Q}_{2t \times 2t} = \begin{pmatrix} 0 & (\mu, \eta) \\ (\eta, \mu) & (\eta, \eta) \end{pmatrix} . \tag{7.19}$$

One can show that the (μ, μ) block always vanishes [163]. Given the structure of the matrix Q_{2t}, the (η, η) block is then irrelevant, since [173]

$$\det \mathbf{Q}_{2t \times 2t} = \big[\det(\mu, \eta) \big]^2 . \tag{7.20}$$

In order to calculate the mixed derivatives in the (μ, η) block (which will be called $\mathbf{B}_{t \times t}$), one needs partial derivatives of the $\bar{\nu}$s with respect to the ηs. These derivatives are obtained by totally differentiating $\partial_{\bar{\nu}_i} \Psi_N = 0$ with respect to the ηs and then setting $\eta_1 = \eta_2 = 0$. The result reads

$$\partial_{\eta_k} \partial_{\mu_l} \Psi_t = 2\mathrm{i} \bigg[\partial_{\bar{\nu}_{l-1}} \partial_{\mu_{l-1}} S_{\sigma,l-1} \delta_{l-1,k} + \Big(\partial^2_{\bar{\nu}_{l-1}} S_{\sigma,l-1} \frac{\partial \bar{\nu}_{l-1}}{\partial \mu_l} + \partial^2_{\mu_l} S_{\sigma,l} \Big) \delta_{l,k} \\ + \partial_{\bar{\nu}_l} \partial_{\mu_l} S_l \frac{\partial \bar{\nu}_l}{\partial \mu_{l+1}} \delta_{l+1,k} \bigg] , \tag{7.21}$$

where $\delta_{l,k}$ is the Kronecker delta.

After the saddle-point integration is done, the expression for the tth trace reads

$$\operatorname{tr} P^t = 2^t \sum_{\text{p.p.}} \left(\prod_{l=1}^{t} \frac{\partial \bar{\nu}_l}{\partial \mu_{l+1}} \bigg|_{\tilde{E}=0} |C(\bar{\nu}_l, \mu_l)|^2 \right) \sqrt{\frac{(2\pi)^{2t}}{J^{2t} |\det \mathbf{B}_{t \times t}|^2}} e^{J \sum_{i=1}^{t} R_i} . \tag{7.22}$$

The sum extends over all periodic points σ of the combined dissipative classical map f^t_{cl}. We need to calculate the determinant of $\mathbf{B}_{t \times t}$, which according to (7.21) is a tridiagonal nonsymmetric matrix with an additional nonzero element in the upper right and lower left corners. Such a determinant can be expressed as the difference between the traces of two different products of 2×2 matrices (see Appendix B). Combining (7.21), (B.2) and (7.22), we are led to

$$\operatorname{tr} P^t = \sum_{\text{p.p.}} \frac{e^{J \sum_{i=1}^t R_i}}{\left| \operatorname{tr} \prod_{l=t}^1 \mathbf{M}_d^{(l)} - \operatorname{tr} \prod_{l=t}^1 \mathbf{M}_l \right|} . \quad (7.23)$$

The inverted order of the indices in the products indicates that the matrices are ordered from left to right according to decreasing indices. The matrix $\mathbf{M}_d^{(l)}$ in the denominator is already the monodromy matrix for the purely dissipative part, $\mathbf{M}_d^{(l)} = \operatorname{diag}(d\mu_d(\bar{\nu}_l)/d\bar{\nu}_l, 1)$. The ordering of the matrix elements is such that the upper left element is $\partial p(p', q')/\partial p'$ and the lower right element is $\partial q(p', q')/\partial q'$. The matrix $\mathbf{M}_l$ is given by

$$\mathbf{M}_l = -\frac{1}{\partial_{\nu_l} \partial_{\mu_l} S_l} \begin{pmatrix} \partial_{\bar{\nu}_l}^2 S_l + (\partial_{\mu_{l+1}}^2 S_{l+1})(d\mu_d(\bar{\nu}_l)/d\bar{\nu}_l) & -(\partial_{\bar{\nu}_l} \partial_{\mu_l} S_l)^2 \\ d\mu_d(\bar{\nu}_l)/d\bar{\nu}_l & 0 \end{pmatrix} ,$$

which unfortunately is not the monodromy matrix $\mathbf{M}^{(l)}$ for the entire lth step. The latter takes the form

$$\mathbf{M}^{(l)} = \frac{1}{\partial_{\nu_l} \partial_{\mu_l} S_l} \begin{pmatrix} -(\partial_{\mu_l}^2 S_l)(d\mu_d(\bar{\nu}_l)/d\bar{\nu}_l) & d\mu_d(\bar{\nu}_l)/d\bar{\nu}_l \\ -(\partial_{\bar{\nu}_l} \partial_{\mu_l} S_l)^2 + (\partial_{\bar{\nu}_l}^2 S_l)(\partial_{\mu_l}^2 S_l) & -\partial_{\bar{\nu}_l}^2 S_l \end{pmatrix} \quad (7.24)$$

when expressed in terms of derivatives of S_l and μ_d. The fact that in $\mathbf{M}_l$ both the indices l and $l+1$ appear makes it impossible to find a similarity transformation independent of l that transforms $\mathbf{M}_l$ to $\mathbf{M}^{(l)}$. Nevertheless, it was shown in [163] that the traces of $\prod_{l=t}^1 \mathbf{M}_l$ and $\prod_{l=t}^1 \mathbf{M}^{(l)}$ are equal. The proof is very technical and I shall not repeat it here.

The last thing to consider is the sign of each saddle-point contribution for arbitrary t. An efficient method of calculating the phase is presented in Appendix A. Luckily, it is not necessary to diagonalize the matrix $\mathbf{Q}_{2t \times 2t}$; a knowledge of all minors of the matrix suffices. I shall show now that it is always possible to choose all minors of $\mathbf{Q}_{2t \times 2t}$ to be real and positive. Equation (A.3) then implies that the sign of each saddle-point contribution in (7.28) is positive.

Observe first of all that, without regularization, all minors D_l of $\mathbf{Q}_{2t \times 2t}$ with the exception of the determinant $\det \mathbf{Q}_{2t \times 2t}$ itself are zero. This is obvious for $l = 1, \ldots, t$, since in that case the corresponding matrix is part of the upper left zero block of $\mathbf{Q}_{2t \times 2t}$. For $l = t + m > t$, note that D_l contains a $t \times t$ upper left block which is zero, and a $t \times m$ (μ, η) block in the upper right corner. Upon expanding D_l, after the first row one encounters subdeterminants with a $(t-1) \times t$ upper left zero block and a $(t-1) \times (m-1)$ upper right (μ, η) block. Both blocks together have $t-1$ rows, in each of which only the $m-1$ elements on the right can be different from zero. Therefore the first $t-1$ rows are always linearly dependent, unless $m = t$, the case that corresponds to the full determinant. Thus, all minors D_l with $1 \leq l \leq 2t-1$ are zero.

Suppose now that we add to Ψ_t a small quadratic term that vanishes at the saddle point $(\boldsymbol{\mu}, \boldsymbol{\eta}) = (\boldsymbol{\mu}_{\mathrm{sp}}, 0)$ and has a maximum there, i.e. a function $-\epsilon \sum_{i=1}^{t} \left((\mu_i - \mu_{\mathrm{sp}_i})^2 + \eta_i^2 \right)$ with infinitesimal $\epsilon > 0$. If the original integral is convergent, the small addition will not change the value of the integral in the limit $\epsilon \to 0$, but it allows us to determine the phase of all minors D_l. The matrices $\mathbf{D}_l$ corresponding to D_l are all replaced by $\mathbf{D}'_l = \mathbf{D}_l + \epsilon \mathbf{1}_l$, where $\mathbf{1}_l$ is the unit matrix in l dimensions. For $1 \leq l \leq t$ we are immediately led to $D_l = \epsilon^l > 0$. For $t+1 \leq l \leq 2t-1$ we expand D'_l in powers of ϵ, and obtain $D'_l = D_l + \epsilon \operatorname{tr} \mathbf{D}_l + \mathcal{O}(\epsilon^2) = \epsilon \operatorname{tr} \mathbf{D}_l + \mathcal{O}(\epsilon^2)$. To determine the traces $\operatorname{tr} \mathbf{D}_l$ we need the second derivatives in the (η, η) block of $\mathbf{Q}_{2t \times 2t}$. They are given by

$$\partial_{\eta_k} \partial_{\eta_l} \Psi_t = \left(\partial^2_{\eta_k} R(\mu_k, \overline{\nu}_{k+1}; \eta_k) + 4 \frac{(\partial_{\overline{\nu}_k} \partial_{\mu_k} S_k)^2}{\partial^2_{\overline{\nu}_k} R(\mu_{k-1}, \overline{\nu}_k; \eta_{k-1})} \right.$$
$$\left. + 4 \frac{(\partial^2_{\overline{\nu}_{k-1}} S_{k-1})^2}{\partial^2_{\overline{\nu}_{k-1}} R(\mu_{k-2}, \overline{\nu}_{k-1}; \eta_{k-2})} \right) \delta_{k,l}$$
$$+ 4 \frac{(\partial^2_{\overline{\nu}_{k-1}} S_{k-1})(\partial_{\overline{\nu}_{k-1}} \partial_{\mu_{k-1}} S_{k-1})}{\partial^2_{\overline{\nu}_{k-1}} R(\mu_{k-2}, \overline{\nu}_{k-1}; \eta_{k-2})} \delta_{k-1,l}$$
$$+ 4 \frac{(\partial_{\overline{\nu}_k} \partial_{\mu_k} S_k)(\partial^2_{\overline{\nu}_k} S_k)}{\partial^2_{\overline{\nu}_k} R(\mu_{k-1}, \overline{\nu}_k; \eta_{k-1})} \delta_{k+1,l} \,.$$

Since $\partial^2_{\eta_k} R(\mu_k, \overline{\nu}_{k+1}; \eta_k) < 0$ and $\partial^2_{\overline{\nu}_k} R(\mu_{k-1}, \overline{\nu}_k; \eta_{k-1}) < 0$ for all k at the saddle point, the diagonal elements of the (η, η) block of $\mathbf{Q}_{2t \times 2t}$ are all real and positive definite (remember that we took out a minus sign in the definition of $\mathbf{Q}_{2t \times 2t}$ in terms of second derivatives). Therefore $\operatorname{tr} \mathbf{D}_l$ is real and larger than zero for all l. Since $\det \mathbf{Q}_{2t \times 2t}$ is also real and positive, the phase of each saddle-point contribution is now determined to be zero, i.e. each saddle point contributes a real, positive number (cf. (A.3)).

With the help of the total monodromy matrix $\mathbf{M} = \prod_{l=t}^{1} \mathbf{M}_l$, the trace can be written as

$$\operatorname{tr} P^t = \sum_{\mathrm{p.p.}} \frac{e^{J \sum_{i=1}^{t} R_i}}{\left| \operatorname{tr} \prod_{i=t}^{1} \mathbf{M}_d^{(i)} - \operatorname{tr} \mathbf{M} \right|} \,. \tag{7.25}$$

Let me massage the expression a bit further. Readers tired of lengthy mathematical manipulations may gain strength by peeking ahead a bit and rejoicing at the beautiful and very simple final result (7.27).

The fact that $\mathbf{M}$ in (7.25) is a 2×2 matrix leads immediately to $\det(\mathbf{1} - \mathbf{M}) = 1 + \det \mathbf{M} - \operatorname{tr} \mathbf{M}$. Since the map is a periodic succession of unitary evolutions (with stability matrices $\mathbf{M}_u^{(i)}$) and dissipative evolutions (with stability matrices $\mathbf{M}_d^{(i)}$), $\mathbf{M}$ is given by the product $\mathbf{M} = \prod_{i=t}^{1} \mathbf{M}_d^{(i)} \mathbf{M}_u^{(i)}$. The stability matrices $\mathbf{M}_u^{(i)}$ are all unitary, so that $\det \mathbf{M}_u^{(i)} = 1$ for all $i = 1 \ldots t$

and $\det \mathbf{M} = \prod_{i=t}^{1} \det \mathbf{M}_{\mathrm{d}}^{(i)}$. As mentioned earlier, the matrix $\mathbf{M}_{\mathrm{d}}^{(i)}$ is diagonal for the dissipative process for which (7.25) was derived;

$$\mathbf{M}_{\mathrm{d}}^{(i)} = \begin{pmatrix} m_{\mathrm{d}}^{(i)} & 0 \\ 0 & 1 \end{pmatrix}. \tag{7.26}$$

But then $\det \mathbf{M}_{\mathrm{d}}^{(i)} = m_{\mathrm{d}}^{(i)}$, and we find

$$\left| \operatorname{tr} \prod_{i=t}^{1} \mathbf{M}_{\mathrm{d}}^{(i)} - \operatorname{tr} \mathbf{M} \right| = \left| 1 + \det \prod_{i=t}^{1} \mathbf{M}_{\mathrm{d}}^{(i)} - \operatorname{tr} \mathbf{M} \right| = |1 + \det \mathbf{M} - \operatorname{tr} \mathbf{M}|$$
$$= |\det(\mathbf{1} - \mathbf{M})|.$$

The actions R_i are zero on the classical trajectories for the dissipative process (4.11), as one can immediately see by using their explicit form [15]. As shown in Chap. 4, the vanishing of the actions can be traced back more generally to conservation of probability by the master equation and therefore holds for other master equations of the same structure as well. But then, *the trace formula (7.28) is identical to the classical trace formula:*

$$\operatorname{tr} P^t = \sum_{\mathrm{p.p.}} \frac{1}{|\det(\mathbf{1} - \mathbf{M})|} \tag{7.27}$$

$$= \sum_{\mathrm{p.o.}} \sum_{r=1}^{\infty} \frac{n_p \delta_{t, n_p r}}{|\det(\mathbf{1} - \mathbf{M}_p^r)|} = \operatorname{tr} P_{\mathrm{cl}}^t, \tag{7.28}$$

where the first sum in (7.28) is over all primitive periodic orbits p of length n_p, r is their repetition number and $\mathbf{M}_p$ their stability matrix. In (7.27), p.p. labels all periodic points belonging to a periodic orbit of total length t, including the repetitions, and $\mathbf{M}$ is the stability matrix for the entire orbit.

The formula (7.27) generalizes Tabor's formula (3.13) for classically area-preserving maps to a non-area-preserving map and shows that even in the case of dissipative quantum maps, information about the spectrum is encoded in classical periodic orbits. All quantities must be evaluated on the periodic orbits. The formula holds for both chaotic and integrable maps, as long as the periodic orbits are sufficiently well separated in phase space so that the SPA is valid.

For comparing (7.27) with Tabor's result, one should remember that we consider here the propagator of the density matrix, but Tabor considers the propagator of the wave function [91]. In the limit of zero dissipation, we should obtain $\operatorname{tr} P = |\operatorname{tr} F|^2$. That limit can unfortunately not be taken in (7.27), since the semiclassical dissipative propagator is valid only for $\tau \gtrsim 1/J$. However, evaluation of $|\operatorname{tr} F|^2$ for $\tau \to 0$ would definitely lead to a double sum over periodic orbits. Of the double sum only a single sum remains in (7.27); all cross terms between different orbits are killed by decoherence. A small amount of dissipation therefore leads naturally to the "diagonal approximation" often used in semiclassical theories. It amounts to neglecting interference terms between different periodic orbits, an approximation that is known to work at most up to the Heisenberg time $t = j$. For larger times correlations between the classical actions become important and the diagonal approximation definitely breaks down. From the nature of the SPA employed in conventional periodic-orbit theory, one would expect the breakdown much earlier, namely at the Ehrenfest time $t = h^{-1} \ln j$. If we assume that the number of periodic points of $\boldsymbol{f}_{\mathrm{cl}}^t$ grows exponentially with t, the typical distance between two periodic points (i.e. two saddles in the SPA!) becomes comparable to $1/j$ and the SPA should break down.

The validity of (7.27) is certainly restricted to times smaller than the Heisenberg time as well. The reason lies in the errors of order $1/J$ made in establishing the semiclassical approximation for the propagator P. Therefore, P^t has an error of the order of t/J and this limits t to $t \ll J$. More severely, the same problem of the increasing density of periodic points arises and would make us expect a breakdown of (7.27) at the Ehrenfest time. So at first sight the introduction of a small amount of dissipation does not extend the validity of the semiclassical approximation, but it allows us to *derive* rigorously the diagonal approximation for times smaller than the Ehrenfest time, for traces of P.

But dissipative systems do have an important advantage over nondissipative ones. The reason is that both the Heisenberg time and the Ehrenfest time play a less crucial role in the presence of dissipation. We shall see in Sect. 7.2.3 that the eigenvalues of P with the largest absolute values converge to the largest Ruelle resonances. The quantum system therefore reaches an invariant state on the *classical* timescale set by the Ruelle resonances. For large J the Heisenberg time is always beyond that classical (J-independent) time; for exponentially large J values the same statement also holds for the Ehrenfest time. The traces of P^t decay converge exponentially to unity for large t, $\mathrm{tr} P^t \to 1$ for $t \to \infty$, as from the definition all eigenvalues besides unity are absolutely smaller than unity. Because the largest eigenvalues converge to the Ruelle resonances for large J, the timescale for the exponential decay is set by these resonances, and by the time the Heisenberg time (or for exponentially large J even the Ehrenfest time) is reached, the traces are just constant and equal to unity up to exponentially small corrections. One may therefore conclude that (7.27) is correct for all *relevant* times. A similar statement about the decay of time dependent expectation values and correlation functions holds, as we shall see.

Tabor's preexponential factor is reproduced in the limit $\tau \to 0$ with the exception of the power. For $\mathbf{M}_\mathrm{d}$ becomes the unit matrix, and thus $\mathbf{M} = \prod_{i=t}^{1} \mathbf{M}_r^{(i)}$. It is, however, raised to the power 1 instead of 1/2, since we propagate a density matrix and not a wave function.

7.2.2 Numerical Verification

Let me dwell a bit more on the precision and range of validity of (7.27). I have calculated numerically the exact quantum mechanical traces for the dissipative kicked top and compared them with the traces obtained from the trace formula (7.27).

The quantum mechanical propagator P is most conveniently calculated in the J_z basis, since the torsion part is then already diagonal. The rotation about the y axis leads to a Wigner d function, whose values we obtained numerically via a recursion relation as described in [92]. The propagator for the dissipation was obtained by inverting numerically the exactly known Laplace image [107, 117]. The total propagator P is a full, complex, non-Hermitian, nonunitary matrix of dimension $(2j+1)^2 \times (2j+1)^2$. Since for the first trace a knowledge of the diagonal matrix elements suffices, I was able to calculate $\text{tr}\,P$ up to $j = 80$. Higher traces are most efficiently obtained via diagonalization, which limited the numerics to $j \leq 40$.

The effort involved in calculating the first classical trace is comparatively small. In all examples considered and even in the presence of a strange attractor, P_cl had at most four fixed points, which could easily be found numerically by a simple Newton method in two dimensions. For each fixed point the stability matrix was found via the formulae given in Appendix C, and so the trace was immediately obtained. But enough of the numerical details – here are the results.

The First Trace

In Fig. 7.1 I show $\text{tr}\,P$ for different values of j as a function of τ and compare it with $\text{tr}\,P_\text{cl}$, (7.27). The values of torsion strength and rotation angle, $k = 4.0$ and $\beta = 2.0$, were chosen such that the system was already rather chaotic in the dissipation-free case at $\tau = 0$ (see Fig. 6.1). With the exception of the case of very small damping, $\text{tr}\,P_\text{cl}$ reproduces $\text{tr}\,P$ perfectly well for all τ, in spite of the strongly changing phase space structure. The agreement extends to smaller τ with increasing j, as is to be expected from the condition of validity of the semiclassical approximation, $\tau \gtrsim 1/J$ [15]. An analysis of the fixed points shows that at $k = 4.0$, $\beta = 2.0$ two fixed points always exist for $\tau \gtrsim 0.1$. Their μ component slowly decreases and the lower fixed point converges towards the south pole with increasing τ, where it becomes a stronger and stronger point attractor.

Figure 7.2 shows the fixed-point structure for a more complicated situation ($k = 8.0$, $\beta = 2.0$). The dissipation-free dynamics at $\tau = 0$ is strongly chaotic; no visible phase space structure is left.

In Fig. 7.3 the first trace is plotted as a function of τ for this situation. The classical trace diverges whenever a bifurcation is reached. Such a behavior is well known from the Gutzwiller formula in the unitary case; the reason

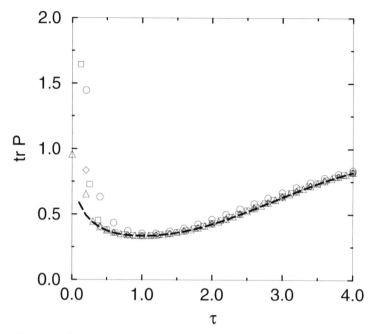

Fig. 7.1. Quantum mechanical traces [$j = 10$ (*circles*), $j = 30$ (*squares*), $j = 50$ (*diamonds*) and $j = 80$ (*triangles*)] as function of dissipation, compared with the classical result (*dashed line*) for $k = 4.0$, $\beta = 2.0$

for the divergence is easily identified as a breakdown of the saddle-point approximation in the semiclassical derivation of the trace formula. While the quantum mechanical traces for small j (say $j \simeq 10$) seem not to take notice of the bifurcations; they approximate the jumps and divergences better and better as j is increased. At $j = 80$ the agreement with the classical trace is very good between the bifurcations. It is remarkable, however, that there are some values of τ close to the bifurcations, where all tr$P$ curves for different j in the entire j range examined cross. The trace seems to be independent of j at these points, but the curves nevertheless do not lie on the classical curve. One is reminded of a Gibbs phenomenon, but I do not have any explanation for it.

Higher Traces

Let us now examine higher traces $\mathrm{tr}\, P^t$ for given values of k, β and τ as a function of t. I shall focus on two limiting cases: the case where the basic phase space structure is a point attractor and the case where it is a well-extended strange attractor. As explained in Chap. 6, a point attractor can always be obtained with sufficiently strong damping.

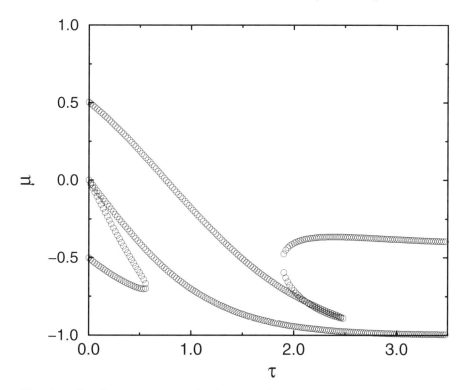

Fig. 7.2. Fixed point structure for $k = 8.0$, $\beta = 2.0$ as a function of τ. The μ component of the fixed points is plotted. There are four fixed points at $\tau = 0.0$, of which two coincide and disappear at $\tau \simeq 0.57$. A new pair is born at $\tau \simeq 1.89$, but one fixed point disappears again at $\tau \simeq 2.47$, in the close vicinity of one of the original fixed points

Consider the example $k = 4.0$, $\beta = 2.0$ and $\tau = 4.0$. Figure 7.4 shows that both the quantum mechanical and the classical result indeed converge rapidly towards unity, and the agreement is very good even for $j = 10$. If one examines the convergence rate one finds that it is slightly j-dependent, but rapidly reaches the classical value (see inset of Fig. 7.4). It should be noted that the calculation of tr P_{cl}^t is simplified enormously here by the fact that with increasing t, no additional periodic points arise. The dissipation is so strong that the system is integrable again. In the example given, there are only two fixed points, one at $(\mu, \phi) \simeq (-0.3812219, -3.098751)$, a strong point repeller, and one at $(\mu, \phi) \simeq (-0.9984018, -1.444154)$, a strong point attractor, and all periodic points of $\boldsymbol{f}_{cl}^t$ are just repetitions of these two points.

The situation is quite different in the case of a strange attractor (Fig. 7.5). Here the number of periodic points increases exponentially with t, as is typical for chaotic systems. This makes the classical calculation of higher traces exceedingly difficult. For $k = 8.0$, $\beta = 2.0$, and $\tau = 1.0$ I was able to

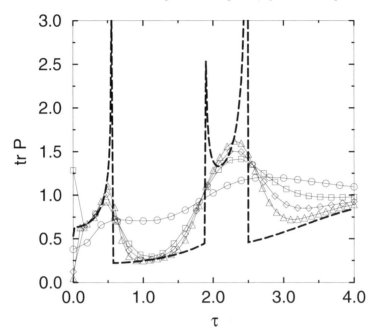

Fig. 7.3. Comparison of quantum mechanical traces with classical trace for $k = 8.0$, $\beta = 2.0$ as a function of τ (same symbols as in Fig. 7.1). The classical trace diverges whenever a bifurcation is reached

calculate tr P_{cl}^t reliably up to $t = 5$, where about 400 periodic points have to be taken into account. The numerical result obtained for tr P_{cl}^t can always be considered as a lower bound on the exact result for tr P_{cl}^t, as long as one can exclude overcounting of fixed points, since all terms in the sum (7.27) are positive. It is then clear that at $t = 5$ the quantum mechanical result for $j = 40$ is still more than a factor 3 away from trP_{cl}^t, even though for $t = 1$ the agreement is very good. The convergence of tr P^t to tr P_{cl}^t as a function of j obviously becomes worse with increasing t.

7.2.3 Leading Eigenvalues

Now that the traces have been calculated, it would appear straightforward to apply Newton's formula in order to obtain the characteristic polynomial and then the eigenvalues of P. In practice, two severe problems arise:

1. To calculate all of the $(2j + 1)^2$ eigenvalues pertaining to the spectrum of P, we would need $t = (2j + 1)^2$ traces. However, in the semiclassical evaluation t is limited to $t \ll 2j + 1$.
2. Even with the "exact" traces, a reconstruction of the entire spectrum of P is prevented by numerical instabilities for $j \geq 3$ (see Fig. 7.6) if

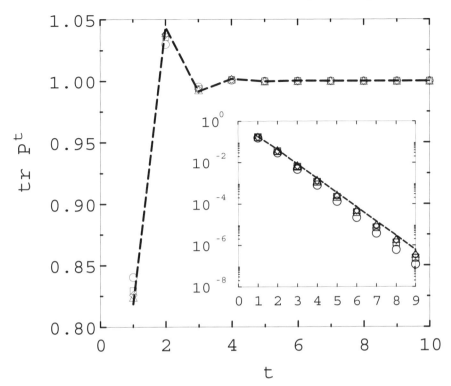

Fig. 7.4. Quantum mechanical and classical traces $\mathrm{tr}\, P^t$ and $\mathrm{tr}\, P_{\mathrm{cl}}^t$ as a function of t for $k = 4.0$, $\beta = 2.0$ and $\tau = 4.0$ ($j = 10$ (*circles*), $j = 20$ (*squares*), $j = 30$ (*diamonds*) and $j = 40$ (*triangles*)). The classical trace is shown as *dashed line* for better visibility, even though it is defined only for integer t. The *inset* shows $|\mathrm{tr}\, P^t - 1|$. So, the exponential convergence to 1 also holds in the classical case. The classical dynamics is dominated here by a single point attractor/repeller pair

some eigenvalues have small absolute value. "Exact" traces means here a finite-precision representation in the computer of the in principle exact values calculated directly from the quantum mechanical eigenvalues of P.

The origin of the instability can be easily identified if we calculate the sensitivity $\partial \lambda_i / \partial t_n$, where t_n is the nth trace $\mathrm{tr}\, P^n$. From the definition of t_n in terms of a sum of nth powers of the eigenvalues, the inverse of the partial derivative is easily obtained, and we find

$$\partial \lambda_i / \partial t_n = 1/(n\lambda_i^{n-1}). \tag{7.29}$$

No problem arises if all eigenvalues are of absolute value unity. But for dissipative quantum maps, many eigenvalues are exponentially small, and this leads to a huge sensitivity of the eigenvalues to small changes in the traces. One can also state this the other way round: since the traces decay expo-

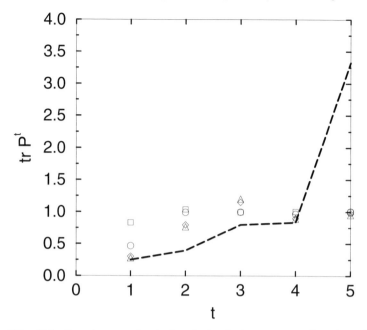

Fig. 7.5. Quantum mechanical and classical traces as a function of t for $k = 8.0$, $\beta = 2.0$ and $\tau = 1.0$ (same symbols as in Fig. 7.4). The corresponding phase space portrait is a strange attractor

nentially towards unity, i.e. $\operatorname{tr} P^t \to 1$ for $t \to \infty$, the higher traces contain hardly any information; the contribution to the traces from all eigenvalues with absolute values smaller than unity decays exponentially with t. Equation (7.29) shows that the smaller an eigenvalue, the more difficult it is to obtain, and the higher a trace, the more sensitively each eigenvalue depends on it.

This observation indicates that the actual problem is not the value of the quantum number j, but rather the number of traces used. The two are not necessarily linked, since we might decide to use a polynomial with a degree smaller than $(2j+1)^2$ if we were not interested in *all* eigenvalues. To see what difference the number of traces used can make, have a look at Figs. 7.7 and 7.8 (yes, do so right now)! In the former, j was chosen as $j = 3$, and 40 and 41 "exact" quantum mechanical traces were fed into Newton's formulae, which were then solved numerically for their roots. The total spectrum comprises 49 eigenvalues. In both cases the eigenvalues with the largest absolute values are still very well reproduced, but the smaller eigenvalues show larger errors. The errors increase substantially from $N_{\max} = 40$ to $N_{\max} = 41$, where $N_{\max}$ denotes the maximum number of traces used, whereas naively one would expect the precision to increase with the number of traces. Upon increasing $N_{\max}$ further the loss of precision increases rapidly, and the eigenvalues found

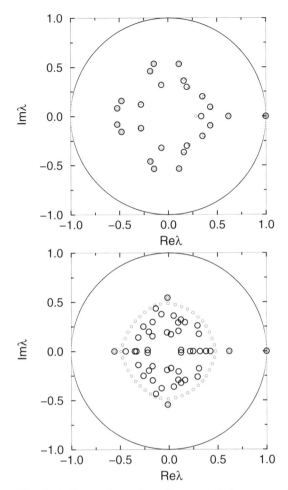

Fig. 7.6. Exact eigenvalues (*large circles*) compared with the eigenvalues obtained from the "exact" traces via Newton's formulae (*small circles*) for $k = 4.0$, $\beta = 2.0$ and $\tau = 0.5$. Up to $j = 2$ (*top*) it is still possible to recover all eigenvalues to good precision, but for larger values of j (e.g. $j = 3$, *bottom*) the numerical instability of the inversion from traces to eigenvalues prevents one finding the eigenvalues even from the numerically "exact" traces

(with the exception of those with an absolute value close to unity) tend to become arranged on a circle within the unit circle. On the other hand, Fig. 7.8 shows that even for $j = 40$, about 20 leading eigenvalues can be reconstructed easily from the "exact" traces if we restrict ourselves to using the first 20 traces.

So, both of the problems mentioned above can be solved at the same time if we restrict ourselves to the *leading* eigenvalues, calculate them from a polynomial of sufficiently low degree and therefore use only traces with

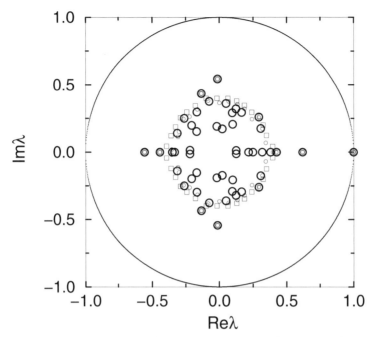

Fig. 7.7. Reconstruction of spectrum for $j = 3$ (*large circles*), using 40 and 41 "exact" traces (*small circles* and *squares*, respectively). In both cases the outer eigenvalues are well reproduced, but the error for the inner ones *increases* substantially if one more trace is included. The system parameters are $k = 4.0$, $\beta = 2.0$ and $\tau = 0.5$

sufficiently small index, which are well described by the semiclassical approximation. Figure 7.9 shows that this works rather well. In the case of strong dissipation, even higher traces could be calculated easily, owing to the simple fixed-point structure. In the example presented, five of the leading eigenvalues are well reproduced by the semiclassical theory, despite the fact that their absolute values are very small.

Figure 7.10 shows the situation for a chaotic case $k = 4$, $\beta = 2$, $\tau = 0.9$. Fixed points up to the sixth iteration could be reliably calculated. The first three quantum mechanical eigenvalues for $j = 40$ are well reproduced.

The restriction to polynomials of sufficiently small degree is also motivated by a zeta function expansion. A zeta function is a characteristic polynomial for any operator A, but in $1/z$ instead of in z: $\zeta(z) = \det(1 - zA)$. Its roots are the inverse eigenvalues of A. It plays an important role in the theory of classical chaotic systems, in which case A is a classical propagator, for example the Frobenius–Perron operator or generalizations of it (see Chap. 2 and Sect. 7.5). Writing the determinant, as in the derivation of Newton's formulae, as $\det(1 - zA) = \exp[\operatorname{tr}\ln(1 - zA)]$ and expanding the logarithm, one is led to

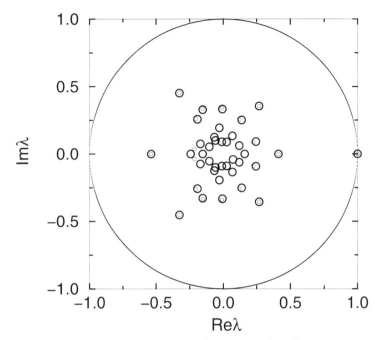

Fig. 7.8. Reconstruction of the leading eigenvalues for $j = 40$, using 20 "exact" traces. The 17 leading eigenvalues can be well reproduced. The *large circles* indicate the exact quantum mechanical eigenvalues ($k = 8.0$, $\beta = 2.0$, $\tau = 1.0$), the *small circles* the reconstructed ones

$$\zeta(z) = \exp\left(1 - \sum_{n=1}^{\infty} \frac{z^n}{n} \operatorname{tr} A^n\right), \tag{7.30}$$

i.e. we have a representation of the zeta function in terms of all traces of A. It is well known that a truncated expansion of the exponential leads to a polynomial with coefficients that decay more rapidly than exponentially, which allows a very accurate extraction of the leading (inverse) eigenvalues of A [174]. If the classical trace formula is inserted in (7.30) the expansion leads to the so-called cycle expansion. This is a sum over pseudo-orbits, where orbits of similar actions are combined. It is known that such a pseudo-cycle expansion can lead to very precise results for the leading eigenvalues of A. But, on the other hand, the expansion of the exponential is exactly what is done in the derivation of Newton's formulae. So, for a finite matrix A, Newton's formulae and the expansion of the zeta function (be it numerically after inserting the traces, or, more sophisticatedly, analytically in the form of the cycle expansion) are completely equivalent – and indeed give the same results with the same precision.

In view of (7.27) one might wonder whether the spectra of P and P_{cl} might not be identical in the limit of $j \to \infty$. Such a conclusion would be a bit

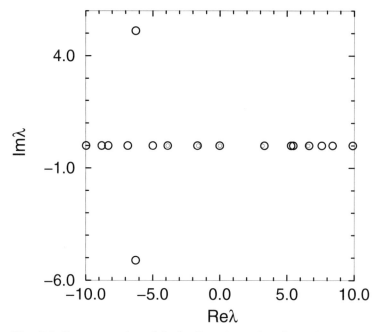

Fig. 7.9. Reconstruction of the leading eigenvalues from the *semiclassically* calculated traces (see (7.27)) in the strongly dissipative case $\tau = 4.0$, $k = 4.0$, $\beta = 2.0$. Five of the leading quantum mechanical eigenvalues for $j = 40$ can be well reproduced. To better display the exponentially small eigenvalues, logarithmic polar coordinates have been used, i.e. $(r, \phi) \to (\ln r, \phi)$. The *large circles* indicate the exact quantum mechanical eigenvalues, the *small circles* the reconstructed ones

premature, however, for at least two reasons. The first one is of formal nature: as mentioned earlier, (7.27) is valid only for $t \ll j$. But in order to determine uniquely the $(2j+1)^2$ eigenvalues, $t = (2j+1)^2$ traces are needed, so that for the highest ones (7.27) is definitely not applicable anymore. Second, as stated earlier in Sect. 2.3.5, $P_{\rm cl}$ necessarily has a continuous spectrum if the system is chaotic. But P has, for all arbitrarily large but finite j, a discrete spectrum, since it is represented by a $(2j+1)^2$-dimensional matrix.

Nevertheless, I would like to argue that *the leading eigenvalues of P coincide with the Ruelle resonances for $j \to \infty$*. This claim is based on the fact that the lowest traces of P and $P_{\rm cl}$ agree for $j \to \infty$ (see (7.27)) and the observation that the first traces determine the leading eigenvalues (see Figs. 7.8, 7.9, and 7.10). Moreover, in Fig. 7.11, where I have plotted the absolutely largest eigenvalues as functions of j, I give direct numerical evidence for the above statement. First, these eigenvalues converge to limiting values independent of j, which must therefore have classical significance. And second, the limiting values coincide with the leading Ruelle resonances ob-

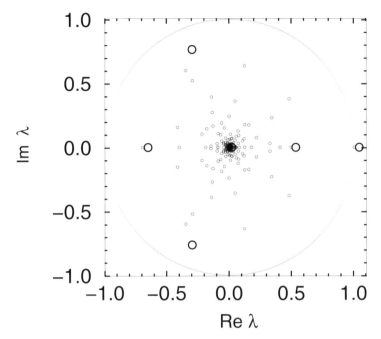

Fig. 7.10. Reconstruction of the leading eigenvalues from the semiclassically calculated traces (see (7.27)) in the chaotic case $k = 4.0$, $\beta = 2.0$, $\tau = 0.9$. The first three leading quantum mechanical eigenvalues (*small circles*) for $j = 40$ are well reproduced by the semiclassical eigenvalues (*large circles*)

tained from the classical trace formula via Newton's formulae (or with a zeta function expansion).

This statement has important consequences. It means that, for large enough j, the quantum mechanical timescales observed, for example in correlation functions, become independent of j and settle down to entirely classical values, namely the timescales set by the Ruelle resonances. That is why the condition $t \ll J$ for the validity of the semiclassical approximation is less severe for dissipative quantum maps than for unitary ones. For larger J the traces have long ago decayed to unity before the condition is violated, as the timescale of the decay is set by the j-independent Ruelle resonances.

Furthermore, we should not be surprised if quantum mechanical correlation functions or time-dependent expectation values decay just like the corresponding classical quantities, as I shall show in Sects. 7.4.3 and 7.4.4.

7.2.4 Comparison with RMT Predictions

Since we can calculate at most the first few eigenvalues from the semiclassically obtained traces, it seems impossible to check the statistical predictions of random-matrix theory (RMT) directly on the level of correlations of the

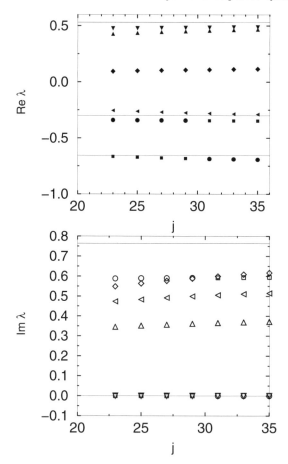

Fig. 7.11. Convergence of the leading eigenvalues of P to the leading Ruelle resonances of P_{cl} as a function of j up to $j = 35$, for $k = 4.0$, $\beta = 2.0$ and $\tau = 0.9$. The real parts of the eigenvalues are plotted with *filled symbols*, the imaginary parts with *open symbols*. Since the eigenvalues are real or come in complex conjugate pairs, only eigenvalues with nonnegative imaginary parts are included. The symbols are: *circles* for the second largest eigenvalue, *squares* for the third, *diamonds* for the fourth, *triangles pointing upwards* for the fifth, *triangles pointing left* for the sixth and *triangles pointing downwards* for the seventh. The second and third eigenvalues switch roles for $j > 29$. The *full lines* indicate the Ruelle resonances obtained from seven classical traces $\mathrm{tr}\, P_{\mathrm{cl}}^t$. The first (the leading one, which is not shown), second and third leading quantum mechanical eigenvalues agree very well with the corresponding Ruelle resonances

eigenvalue density. Instead, it appears more reasonable to take a direct look at the RMT predictions of the traces and compare these predictions with semiclassical or numerical results.

As we have seen in Sect. 7.2.3, the traces of any propagator of a dissipative quantum map must converge to unity, i.e. $\text{tr} P^t \to 1$ for $t \to \infty$. However, from the symmetry of the eigenvalue distribution it is clear that in the Ginibre ensemble $\langle \text{tr} P^t \rangle = 0$ for all t. One might object that the uniform distribution of eigenvalues in the unit circle is a feature of RMT that need not be universal, i.e. a real physical system will typically not have the density of states prescribed by random-matrix theory. Moreover, $\text{tr} P^t$ converges to unity because the role of the eigenvalue $\lambda = 1$ is singled out for the dissipative quantum map, and Ginibre's ensemble does not know about this particular eigenvalue. However, the correlations between eigenvalues should be universal. In nondissipative quantum maps the correlations between eigenvalues are reflected in the average of the *squared absolute* traces, which give directly the spectral form factor. For dissipative quantum maps there is no such direct connection, but nevertheless it clear that the squared absolute traces, $\langle |\text{tr} P^t|^2 \rangle$, do say something about spectral correlations. For

$$\langle |\text{tr} P^t|^2 \rangle = \left\langle \sum_{l,m=1}^{N} z_l^t (z_m^*)^t \right\rangle, \quad (7.31)$$

and if the eigenvalues z_i were uncorrelated, the mixed terms ($l \neq m$) in the expansion of the double sum could be factorized, $\langle z_l^t (z_m^*)^t \rangle = \langle z_l^t \rangle \langle (z_m^*)^t \rangle$. Since the averages equal zero we would obtain $\langle |\text{tr} P^t|^2 \rangle = N \langle |z_1|^{2t} \rangle$. Deviations from this result signal spectral correlations, and one might indeed hope that these correlations were universal. Are they?

Here comes the calculation! I start with a representation of the joint distribution (6.15) in terms of integrals over anticommuting Grassmann variables η_i ($\eta_i \eta_j = -\eta_j \eta_i$) [60],

$$P_N(z_1, \ldots, z_N) \quad (7.32)$$

$$= \left(\prod_{k=1}^{N} \frac{e^{-|z_k|^2}}{\pi k!} \right) \int \left(\prod_{j=1}^{N} d\eta_j^* d\eta_j \right) \prod_{j=1}^{N} \left(-\sum_{ik} \eta_i^* \eta_k z_j^{i-1} (z_j^*)^{k-1} \right).$$

This representation has the advantage that instead of the product of terms $|z_i - z_j|$ we now have an expanded form, where the z_is and z_j^*s appear directly and can be integrated over more easily. The integration measure will be denoted by $d^2 z_i = dx\, dy$ for the decomposition of $z_i = x + iy$ into real and imaginary parts. If we insert (7.32) into the definition of $\langle |\text{tr} P^t|^2 \rangle$,

$$\langle |\text{tr} P^t|^2 \rangle = \int d^2 z_1 \ldots d^2 z_N \sum_{l,m=1}^{N} z_l^t (z_m^*)^t P_N(z_1, \ldots, z_N), \quad (7.33)$$

and separate the diagonal terms in the double sum from the off-diagonal ones, we obtain

$$\langle |\text{tr} P^t|^2 \rangle = \sum_l S_l + \sum_{l \neq m} S_{lm} \quad (7.34)$$

with

$$S_l = \left(\int \prod_j \frac{\mathrm{d}\eta_j^* \mathrm{d}\eta_j}{j!}\right) \int \mathrm{d}^2 z_l \frac{\mathrm{e}^{-|z_l|^2}}{\pi} \left(-\sum_{ik} \eta_i^* \eta_k z_l^{i+t-1}(z_l^*)^{k+t-1}\right)$$
$$\times \int \prod_{j \neq l}^N \mathrm{d}^2 z_j \frac{\mathrm{e}^{-|z_j|^2}}{\pi} \left(-\sum_{ik} \eta_i^* \eta_k z_j^{i-1}(z_j^*)^{k-1}\right), \quad (7.35)$$

and correspondingly

$$S_{lm} = \int \left(\prod_j \frac{\mathrm{d}\eta_j^* \mathrm{d}\eta_j}{j!}\right) \int \mathrm{d}^2 z_l \frac{\mathrm{e}^{-|z_l|^2}}{\pi} \left(-\sum_{ik} \eta_i^* \eta_k z_l^{i+t-1}(z_l^*)^{k-1}\right)$$
$$\times \int \mathrm{d}^2 z_m \frac{\mathrm{e}^{-|z_m|^2}}{\pi} \left(-\sum_{ik} \eta_i^* \eta_k z_m^{i-1}(z_m^*)^{k+t-1}\right)$$
$$\times \int \prod_{j \neq l,m}^N \mathrm{d}^2 z_j \frac{\mathrm{e}^{-|z_j|^2}}{\pi} \left(-\sum_{ik} \eta_i^* \eta_k z_j^{i-1}(z_j^*)^{k-1}\right). \quad (7.36)$$

The integration over the zs can now be performed. One can easily check that

$$\int \mathrm{d}^2 z \frac{\mathrm{e}^{-|z_l|^2}}{\pi} z^m (z^*)^n = m \delta_{m,n} \quad (7.37)$$

for all natural numbers m, n. With the help of the Kronecker delta $\delta_{m,n}$, one of the summations over i, k in (7.35) and (7.36) can be performed. We obtain for S_l

$$S_l = \int \left(\prod_j \frac{\mathrm{d}\eta_j^* \mathrm{d}\eta_j}{j!}\right) \left(-\sum_{i=1}^N \eta_i^* \eta_i (i+t-1)!\right)$$
$$\times \left(-\sum_{k=1}^N \eta_k^* \eta_k (k-1)!\right)^{N-1}.$$

Owing to the anticommuting property of the Grassmann variables and the definition of Grassmann integration, only terms that contain all of the ηs and η^*s exactly once contribute. Out of the $N-1$ factors $\eta_k^* \eta_k$, $(N-1)!$ combinations arise, each of which has to be combined with one term $\eta_i^* \eta_i$ with a different index i. Groups of $\eta_k^* \eta_k$ commute, and the rules of integration for Grassmann variables [60] lead to

$$\int \mathrm{d}\eta^* \mathrm{d}\eta \eta^* \eta = -1. \quad (7.38)$$

With all this, we readily obtain

$$S_l = \frac{1}{N} \sum_{i=1}^{N} \frac{(i+t-1)!}{(i-1)!}. \tag{7.39}$$

The expression for S_{lm} after the z integrations are performed reads

$$S_{lm} = \int \left(\prod_j \frac{\mathrm{d}\eta_j^* \mathrm{d}\eta_j}{j!} \right) \left(-\sum_{i=1}^{N-t} \eta_i^* \eta_{i+t}(i+t-1)! \right)$$

$$\times \left(-\sum_{k=1}^{N-t} \eta_{k+t}^* \eta_k (k+t-1)! \right) \left(-\sum_{n=1}^{N} \eta_n^* \eta_n (n-1)! \right)^{N-2}. \tag{7.40}$$

The Grassmann integration rules imply that in the first two factors i must equal k, and then $n \neq i, i+t$ in all the other terms. Let us assume that $t > 0$. There are $N-2$ factors containing $\eta_n^* \eta_n$, each containing N summands, out of which $N-2$ may be used, and all of them must be different. So this leads to $(N-2)(N-3)\ldots(N-(N-1)) = (N-2)!$ choices from the last $N-2$ factors. The different grouping of the Grassmann variables leads now to an additional factor -1. Altogether, S_{lm} is given by

$$S_{lm} = -\frac{1}{N(N-1)} \sum_{i=1}^{N-t} \frac{(i+t-1)!}{(i-1)!}. \tag{7.41}$$

So both S_l and S_{lm} are independent of their indices, as was of course to be expected, since no eigenvalue is singled out from the others. In the result for the trace the prefactors $1/N$ and $1/N(N-1)$ therefore cancel;

$$\langle |\operatorname{tr} P^t|^2 \rangle = N S_l + N(N-1) S_{lm}$$

$$= \sum_{i=1}^{N} \frac{(i+t-1)!}{(i-1)!} - \sum_{i=1}^{N-t} \frac{(i+t-1)!}{(i-1)!}$$

$$= \sum_{l=\max(0,N-t)}^{N-1} \frac{(l+t)!}{l!}. \tag{7.42}$$

This result is valid for $t \geq 1$. For $t = 0$ the combinatorics in (7.40) are different and lead directly to $S_{lm} = S_l = 1$, as they should. Note that the term NS_l is just the diagonal term, which would be the only one left if there were no correlations. So the presence of the additional term $N(N-1)S_{lm}$ signals spectral correlations. Interestingly enough, this term vanishes for $t > N$, which might be interpreted as effective randomization of the phases of the z_i^t. Even if the phases of the z_i are correlated, taking the tth power increases the phase differences sufficiently that the correlations are lost.

The summation in (7.42) can be performed with the help of the identity [175]

$$\sum_{k=0}^{n} \frac{(k+r)!}{k!} = \frac{(r+n+1)!}{n!(r+1)}, \tag{7.43}$$

and we obtain the final result,

$$\langle |\text{tr}\, P^t|^2 \rangle = \frac{(t+N)!}{(N-1)!(t+1)} - \frac{[t+\max(1, N-t)]!}{\max(0, N-t-1)!(t+1)}. \quad (7.44)$$

In order to compare this result with the traces of a dissipative quantum map, we should renormalize again according to $z = \zeta \sqrt{N}$. This leads to an additional prefactor $1/N^t$, and we have

$$\langle |\text{tr}\, P_\zeta^t|^2 \rangle = \frac{1}{(t+1)N^t} \left(\frac{(t+N)!}{(N-1)!} - \frac{[t+\max(1, N-t)]!}{\max(0, N-t-1)!} \right), \quad (7.45)$$

where P_ζ denotes the renormalized matrix. The asymptotic behavior can be unraveled by use of Stirling's formula. For $N \gg t^2$ one obtains the simple and universal law

$$\langle |\text{tr}\, P_\zeta^t|^2 \rangle \simeq t. \quad (7.46)$$

Quite surprisingly, this is identical to the universal small-time behavior of the CUE form factor! Unfortunately, I have seen no clear indication of this universal short-time behavior in the numerically or semiclassically calculated traces, even after carefully unfolding the spectra. The observed signal is very noisy, and presumably averaging over many systems is required before the universal linear behavior is observed. Semiclassically, one would like to use a kind of Hannay–Ozorio de Almeida sum rule [176], but a generalization to strange attractors would first have to be established. For $t \gg N \gg 1$, the traces behave as

$$\langle |\text{tr}\, P_\zeta^t|^2 \rangle \simeq \frac{e^N}{N^{N-1/2}} t^{N+t-1/2} e^{-t(1+\ln N)}. \quad (7.47)$$

So the traces decay exponentially for sufficiently large t with a rate that depends weakly on N.

7.3 The Wigner Function and its Propagator

In this section I derive a key result that will greatly simplify the semiclassical study of the remaining quantities of interest in the quantum mechanical problem. Since we expect that, in general, decoherence renders the dynamics more classical, it is useful to look from the very beginning for a formulation that is as close as possible to a classical phase space formulation. It is therefore natural to go over to a phase space representation of the density matrix. The Wigner function is very well suited for this purpose. In fact, the Wigner function has been used many times in order to study the transition from quantum to classical mechanics [125, 127, 128, 130, 154, 155, 157, 177, 178]. There are other phase space functions related to the density matrix, such as the Husimi function (also called the Q function in quantum optics), which has properties even closer to a classical phase space density. But the Wigner

7.3 The Wigner Function and its Propagator

function turns out to be entirely sufficient for our purpose. I shall show in this section that the propagator of the Wigner function is – for sufficiently smooth Wigner functions – nothing but the classical Frobenius–Perron propagator of phase space density.

Let me first define the Wigner function, adapted to the present problem. Usually, the Wigner transform is defined as a Fourier transform with respect to the skewness of the density matrix in the coordinate representation, for a flat phase space (see T. Dittrich in [42] and [179, 180]);

$$\rho_W(p,q) = \frac{1}{2\pi\hbar}\int_{-\infty}^{\infty} dx\, e^{ipx/\hbar}\left\langle q-\frac{x}{2}\right|\hat{\rho}\left|q+\frac{x}{2}\right\rangle. \quad (7.48)$$

In our problem, we have ρ in the momentum basis μ and a phase space with the topology of a sphere. Inserting resolutions of the identity operator in the momentum basis into the above definition (7.48) of $\rho_W(p,q)$, we obtain

$$\rho_W(p,q) = \frac{1}{2\pi\hbar}\int_{-\infty}^{\infty} d\xi\, e^{iq\xi/\hbar}\left\langle p+\frac{\xi}{2}\right|\hat{\rho}\left|p-\frac{\xi}{2}\right\rangle. \quad (7.49)$$

Note the change of sign in the skewness. For our spin dynamics we have $p = \mu$, $q = \phi$, and $1/J$ replaces $\hbar$. An additional factor J arises because the original quantum numbers m and k are rescaled to μ and η as explained above. I therefore define the Wigner transform of $\rho(\mu,\eta,t)$ as

$$\rho_W(\mu,\phi,t) = \frac{J^2}{2\pi}\int_{-\infty}^{\infty} d\eta\, e^{iJ\eta\phi}\rho\left(\mu,\frac{\eta}{2},t\right). \quad (7.50)$$

It has the right normalization, in the sense that

$$\int d\mu\, d\phi\, \rho_W(\mu,\phi,t) = J\int d\mu\, \rho(\mu,0,t) \simeq \sum_m \rho_{mm}(t) = 1. \quad (7.51)$$

The corrections from passing from the integral to the discrete sum of the diagonal matrix elements are of order $1/J$ and become negligible in the limit of large J, as long as $\rho(\mu,0,t)$ does not fluctuate on a scale of $1/J$, i.e. as long as the probability profile has a classical meaning.

The inverse transformation reads

$$\rho(\mu,\eta,t) = \frac{1}{J}\int d\phi\, e^{-i2J\eta\phi}\rho_W(\mu,\phi,t). \quad (7.52)$$

Wigner functions on $SU(2)$ have been introduced before in the literature [181, 182, 183, 184, 185] via angular-momentum coherent states and appropriate transformations of Q or P functions. These definitions avoid problems at the poles of the Bloch sphere that can arise in the present approach. On the other hand, the definition (7.50) is much simpler from a technical point of view and sufficient for our purposes.

We are now in a position to calculate the Wigner function obtained after one application of the map, from the original function. We insert the propagated density matrix $\rho(\mu,\eta,t+1)$ from (7.3) into

$$\rho_W(\mu, \phi, t+1) = \frac{J^2}{\pi} \int_{-\infty}^{\infty} d\eta\, e^{2iJ\eta\phi} \rho(\mu, \eta, t+1) \tag{7.53}$$

and then express the original density matrix $\rho(\mu', \eta', t)$ in terms of its Wigner transform;

$$\rho_W(\mu, \phi, t+1) = \frac{2J^3}{\pi} \int d\eta\, d\mu'\, d\eta'\, d\phi' \sum_{s_1, s_2} P(\mu, \eta; \mu', \eta') \rho_W(\mu', \phi', t)$$
$$\times \exp\Big(2iJ\big(\pi((s_1+s_2)\mu' + (s_1-s_2)\eta') - \eta'\phi' + \eta\phi\big)\Big). \tag{7.54}$$

The integers s_1 and s_2 arise from the Poisson summation that transforms the summation over m' and k' into integrals over μ' and η'. With the semiclassical expression (7.11) for the propagator, we arrive at

$$\rho_W(\mu, \phi, t+1)$$
$$= \frac{2J^3}{\pi} \int d\eta\, d\mu'\, d\eta'\, d\phi' \sum_{\sigma_1, \sigma_2 s_1, s_2 \bar{\nu}, l} \exp\big(J\Psi(\mu, \eta; \mu', \eta')\big)$$
$$\times \sqrt{\left(\frac{\partial \bar{\nu}}{\partial \mu}\right)_{\tilde{E}} \left(\frac{\partial \bar{\nu}}{\partial \mu}\right)} C_{\sigma_1}(\bar{\nu}+\eta, \mu'+\eta') C^*_{\sigma_2}(\bar{\nu}-\eta, \mu'-\eta') \rho_W(\mu', \phi', t),$$

where the "action" Ψ is given by

$$\Psi(\mu, \eta; \mu', \eta') \tag{7.55}$$
$$= \psi(\mu, \eta; \mu', \eta') + i\Big(2\eta\phi - 2\eta'\phi' + 2\pi\big[s_1(\mu'+\eta') + s_2(\mu'-\eta')\big]\Big).$$

The form of the integrands and the fact that P is already approximated semiclassically, i.e. correct only to order $1/J$, suggests that we should integrate again using the SPA. To do so, we must assume that the initial Wigner function $\rho_W(\mu', \phi')$ is sufficiently smooth, i.e. has no structure on a scale of $1/J$. This is, at the same time, a necessary condition if we want to attribute a classical meaning to ρ_W.

The saddle-point equations read

$$\partial_\eta \Psi = \partial_\eta R + \partial_{\bar{\nu}} R \partial_\eta \bar{\nu} \tag{7.56}$$
$$+ i\Big(\big(-\phi^f_{\sigma_1} + \phi^f_{\sigma_2} + 2\pi l\big)\partial_\eta \bar{\nu} - \phi^f_{\sigma_1} - \phi^f_{\sigma_2} + 2\phi\Big) = 0,$$
$$\partial_{\mu'} \Psi = \partial_{\bar{\nu}} R \partial_{\mu'} \bar{\nu} \tag{7.57}$$
$$+ i\Big(\big(-\phi^f_{\sigma_1} + \phi^f_{\sigma_2} + 2\pi l\big)\partial_{\mu'}\bar{\nu} + \phi^i_{\sigma_1} - \phi^i_{\sigma_2} + 2\pi(s_1+s_2)\Big) = 0,$$
$$\partial_{\eta'} \Psi = \partial_{\bar{\nu}} R \partial_{\eta'} \bar{\nu}$$
$$+ i\Big(\big(-\phi^f_{\sigma_1} + \phi^f_{\sigma_2} + 2\pi l\big)\partial_{\eta'}\bar{\nu} + \phi^i_{\sigma_1} + \phi^i_{\sigma_2} + 2\pi(s_1-s_2) - 2\phi'\Big)$$
$$= 0, \tag{7.58}$$
$$\partial_{\phi'} \Psi = 2i\eta' = 0. \tag{7.59}$$

For brevity I have suppressed the arguments of R and $\phi^i_\sigma, \phi^f_\sigma$. They are the same as those in (7.8) for R and S_σ, $\sigma = \sigma_1, \sigma_2$. Equation (7.59) immediately gives $\eta' = 0$. To solve the rest of the equations, let us first assume that $\partial_{\bar{\nu}} R = 0$. I shall show below that this is the only possible choice. Then we have, from the general properties of R (see Sect. 4.4.8), the result that μ is connected to $\bar{\nu}$ via the classical dissipative trajectory, $\mu = \mu_d(\bar{\nu})$, and the real parts of (7.57) and (7.58) give zero. As in the calculation of the traces of P, I shall assume that all relevant solutions to the saddle-point equations are real, as is expected from the physical origin of the variables as real-valued quantum numbers. Again, I cannot exclude formally the existence of complex solutions, but as long as classical solutions exist we expect them to dominate over nonclassical ones. This is certainly true in the case of superradiance dissipation, where $R = 0$ on the classical trajectory, whereas complex solutions would lead to exponential suppression. But, even more generally, we expect classical solutions to dominate, since they are known to dominate in nondissipative quantum mechanics and dissipation favors classical behavior even more.

The real and imaginary parts of all saddle-point equations must then separately equal zero, so that we have eight instead of four equations. The assumption $\partial_{\bar{\nu}} R = 0$ solves two of them at the same time. The real part of (7.56) gives additionally $\partial_\eta R = 0$ and thus, according to the general properties of R, $\eta = 0$. Only the propagation of probabilities, i.e. the diagonal elements of the density matrix, contributes in the saddle point approximation.

It follows from (7.9) that $\partial_{\bar{\nu}} R + i\left(-\phi^f_{\sigma_1} + \phi^f_{\sigma_2} + 2\pi l\right) = 0$, i.e. $-\phi^f_{\sigma_1}(\bar{\nu}, \mu) + \phi^f_{\sigma_2}(\bar{\nu}, \mu) + 2\pi l = 0$. Thus, the final canonical coordinates of the two trajectories σ_1 and σ_2 must agree up to integer multiples of 2π. Since the initial and final momenta (μ and $\bar{\nu}$, respectively) are also the same, the two trajectories must be identical, i.e. $\sigma_1 = \sigma_2 \equiv \sigma$. If all angles are counted modulo 2π, we also have $l = 0$.

The imaginary part of (7.56) leads to $\phi^f_\sigma = \phi$, the imaginary part of (7.57) to $s_1 + s_2 = 0$, and the imaginary part of (7.58) to $\phi^i_\sigma = \phi' + 2\pi s_2$. These equations describe precisely the classical trajectories for the unitary part of the motion from an initial phase space point (μ', ϕ') to a final point $(\bar{\nu}, \phi)$, again counting the angles modulo 2π. Together with $\mu = \mu_d(\bar{\nu})$, the saddle-point equations thus give the classical trajectory from (μ', ϕ') to (μ, ϕ). Note that this trajectory is unique if it exists, since classical trajectories are uniquely defined by their starting point in phase space.

For evaluating the SPA we need in addition the determinant of the matrix $\Psi^{(2)}$ of second derivatives of Ψ. It is straightforward to verify that its absolute value is given by

$$|\det \Psi^{(2)}| = 16|\partial_{\bar{\nu}} \phi^i_\sigma(\bar{\nu}, \mu')|^2 . \tag{7.60}$$

The overall phase arising from the SPA equals zero, as can be seen by employing the same techniques as were used for the calculation of $\operatorname{tr} P^t$ (see Sect. 7.2.1). We arrive at the saddle-point approximation

104   7. Semiclassical Analysis of Dissipative Quantum Maps

$$\rho_W(\mu, \phi, t+1) = \frac{8\pi J}{\sqrt{|\det \Psi^{(2)}|}} \left|\frac{\partial \bar{\nu}}{\partial \mu}\right| |C(\bar{\nu}, \mu')|^2 \rho_W\left(\boldsymbol{f}_{\text{cl}}^{-1}(\mu, \phi), t\right)$$

$$= \left|\frac{\partial \bar{\nu}}{\partial \mu}\right| \rho_W\left(\boldsymbol{f}_{\text{cl}}^{-1}(\mu, \phi), t\right). \tag{7.61}$$

The prefactor in the last equation is nothing but the inverse of the Jacobian of the classical trajectory, which arises solely from the dissipative step since the unitary step has a Jacobian of unity. So, with the abbreviations $\boldsymbol{y} = (\mu, \phi)$ and $\boldsymbol{x} = (\mu', \phi')$ for the final and initial phase space coordinates, we have

$$\rho_W(\boldsymbol{y}, t+1) = \frac{\rho_W\left(\boldsymbol{f}_{\text{cl}}^{-1}(\boldsymbol{y}), t\right)}{|\partial \boldsymbol{f}_{\text{cl}}/\partial \boldsymbol{x}|_{\boldsymbol{x}=\boldsymbol{f}_{\text{cl}}^{-1}(\boldsymbol{y})}} = \int d\boldsymbol{x}\, \delta\big(\boldsymbol{y} - \boldsymbol{f}_{\text{cl}}(\boldsymbol{x})\big) \rho_W(\boldsymbol{x}, t)$$

$$\equiv \int d\boldsymbol{x}\, P_W(\boldsymbol{y}, \boldsymbol{x}) \rho_W(\boldsymbol{x}, t). \tag{7.62}$$

This identifies the propagator of the Wigner function as the classical Frobenius–Perron propagator of phase space density,

$$P_W(\boldsymbol{y}, \boldsymbol{x}) = P_{\text{cl}}(\boldsymbol{y}, \boldsymbol{x}) = \delta\big(\boldsymbol{y} - \boldsymbol{f}_{\text{cl}}(\boldsymbol{x})\big). \tag{7.63}$$

Note once more that this conclusion holds only if the test function $\rho_W(x)$ on which P_W acts is sufficiently smooth, namely if it does not contain any structure on a scale of $1/J$ or smaller. For classical phase space densities this is often not the case. Indeed, continued application of a chaotic map leads to ever finer phase space structure, so that after the Ehrenfest time, of order $\ln J$, scales are reached that are comparable with $\hbar$. That is why the validity of the equation $P_W^t = P_{\text{cl}}^t$ which follows immediately from equation (7.63) is restricted to discrete times t smaller than the Ehrenfest time.

Let me show, finally, that there is no alternative to the assumption $\partial_{\bar{\nu}} R = 0$ about the solution of the saddle-point equation if only real solutions are permitted. To see this, suppose that $\partial_{\bar{\nu}} R \neq 0$. Then we have from (7.57) that $\partial \bar{\nu}/\partial \mu' = 0$, and from (7.58) that $\partial \bar{\nu}/\partial \eta' = 0$, such that $\bar{\nu}$ is a function of μ and η alone. The imaginary part of (7.57) gives $\phi_{\sigma_1}^i - \phi_{\sigma_2}^i + 2\pi(s_1 + s_2) = 0$. If we differentiate with respect to μ' and η' and remember that $\eta' = 0$ follows directly from (7.59), we are immediately led to $\partial_{\mu'} \phi_{\sigma_1}^i(\bar{\nu} + \eta, \mu') = \partial_{\mu'} \phi_{\sigma_2}^i(\bar{\nu} - \eta, \mu') = 0$. Thus, all trajectories with a given initial ϕ_σ^i end at the same final momentum $\bar{\nu} + \eta$ (for $\sigma = \sigma_1$) or $\bar{\nu} - \eta$ (for $\sigma = \sigma_2$). From the imaginary part of (7.56), it follows in the same fashion that $\partial_{\mu'} \phi_{\sigma_1}^f(\bar{\nu}+\eta, \mu') = \partial_{\mu'} \phi_{\sigma_2}^f(\bar{\nu} - \eta, \mu') = 0$. So the final canonical coordinate does not depend on the initial momentum either. In other words, all trajectories with the same initial $\phi_{\sigma_1}^i$ (and correspondingly, for the same initial $\phi_{\sigma_2}^i$) but arbitrary initial μ' end at the same final phase space point. But this is in contradiction to the fact that a final phase space point uniquely defines a trajectory. Therefore, the initial assumption $\partial_{\bar{\nu}} R \neq 0$ must be wrong, QED.

We conclude that *the Wigner propagator is, up to the Ehrenfest time, just the classical propagator*. It is not important whether the initial Wigner function is a classical phase space density or not, as long as it has no structure on the scale of $1/J$ or smaller. Numerical investigations show that even in the case of rapid oscillations of ρ_W, the results are not too bad, though. Figure 7.12 demonstrates the "worst-case scenario", where I have propagated the Wigner function for an initial Schrödinger cat state using the exact quantum mechanical propagator and using the classical propagator. The classical propagation preserves some of the wiggles in the original Wigner function which are smeared out by the exact propagation, but the overall shape is well reproduced. Similar conclusions about the validity of the classical propagator for the initial evolution of the Wigner function have also been reached by other authors [125, 127, 130, 154].

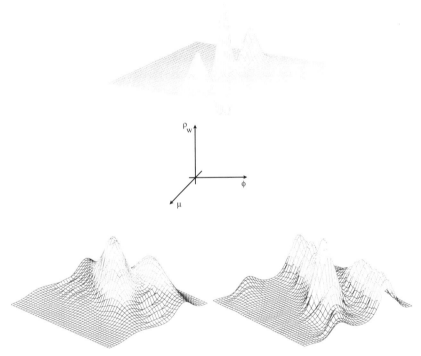

Fig. 7.12. Comparison of the propagation of a Schrödinger cat state (*top*) using the exact quantum mechanical propagator (*left*) and using the classical propagator (*right*) after two iterations. The initial Schrödinger cat state (*top*) consists of a superposition of two angular-momentum coherent states with $\gamma_1 = 0.5$ and $\gamma_2 = 2.0$. After two iterations the Wigner function is positive nearly everywhere, and the classically propagated function looks very similiar to the one propagated using the exact quantum mechanical propagator

7.4 Consequences

7.4.1 The Trace Formula Revisited

Equation (7.63) allows very easily to verify the trace formula (7.27) derived in Sect. 7.2.1. All that remains to be done is to verify that $\operatorname{tr} P^t = \operatorname{tr} P_W^t$. To see that this is the case, let us extract the general relation between any P_W and the corresponding P from (7.54) and the definition of P_W in (7.62). Comparing the two equations, we are led to

$$P_W(\mu, \phi; \mu', \phi') = \frac{2J^3}{\pi} \int d\eta \, d\eta' \sum_{s_1, s_2} e^{2iJ(\eta'\phi' - \eta\phi) + i2\pi J\big((s_1+s_2)\mu + (s_1-s_2)\eta'\big)} P(\mu, \eta; \mu', \eta').$$

This equation holds for any propagator P of the density matrix, and therefore also for the propagator P^t of the tth iteration of the original map. It is then one line of calculation to show that the trace of P_W^t,

$$\operatorname{tr} P_W^t = \int d\mu \, d\phi \, P_W^t(\mu, \phi; \mu, \phi), \tag{7.64}$$

is given by

$$\operatorname{tr} P_W^t = 2J^2 \int d\eta \, d\mu \sum_{s_1, s_2} e^{i2\pi J[(s_1+s_2)\mu + (s_1-s_2)\eta]} P^t(\mu, \eta; \mu, \eta)$$

$$= \operatorname{tr} P^t, \tag{7.65}$$

where in the last equation I have gone back to discrete summation, undoing the Poisson summation. Thus, we have, up to $\mathcal{O}(1/J)$ corrections, $\operatorname{tr} P^t = \operatorname{tr} P_W^t = \operatorname{tr} P_{cl}^t$, QED.

7.4.2 The Invariant State

If the classical and the Wigner propagator are the same up to corrections of order $1/J$, so are their eigenstates. An invariant state of P is defined as an eigenstate with eigenvalue one, i.e. $P\rho(\infty) = \rho(\infty)$, and correspondingly $P_W \rho_W(\infty) = \rho_W(\infty)$. Without special symmetries, this eigenvalue is nondegenerate. As in the classical case (see Chap. 2), I indicate by the arguments "infinity" that the invariant state can also be reached by iterating the map infinitely many times. The classical ergodic state ($P_{cl}\rho_{cl}(\infty) = \rho_{cl}(\infty)$) has its quantum mechanical correspondence, as we shall see presently.

From (7.63) we conclude, that up to $\mathcal{O}(1/J)$ corrections,

$$\rho_W(\infty) = \rho_{cl}(\infty). \tag{7.66}$$

The corrections have to be understood as a smearing out on a scale of $1/J$. Indeed, suppose we start from a smooth initial Wigner function and then iterate it many times with P_W; it evolves according to (7.63) up to the Ehrenfest time as a classical phase space density. After the Ehrenfest time the classical dynamics continues to produce ever finer structures in the phase space density, whereas Heisenberg's uncertainty principle prohibits structures in ρ_W smaller than $1/J$. As pointed out before, (7.63) therefore ceases to be valid, and ρ_W is left at the stage where it is the smeared-out classical strange attractor. Figure 7.13 shows that indeed the quantum strange attractor is a smeared-out classical one. The Wigner function was obtained by direct diagonalization of the propagator and subsequent Wigner transformation of the eigenstate with eigenvalue unity, the classical picture by iterating a classical

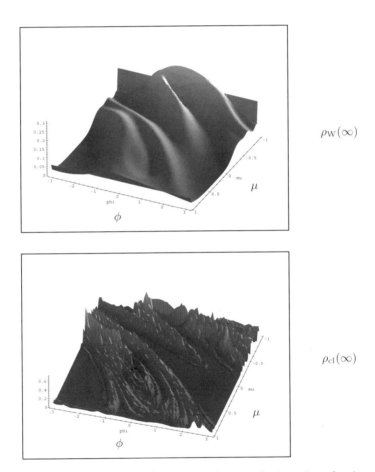

Fig. 7.13. The Wigner function $\rho_W(\infty)$ corresponding to the invariant density matrix (for $j = 40$), and the classical stationary probability distribution $\rho_{cl}(\infty)$ (strange attractor) for $k = 4.0$, $\beta = 2.0$ and $\tau = 0.5$. The Wigner function is a "quantum strange attractor", a smeared-out version of the classical strange attractor. The range of arguments is $\phi = -\pi$ to π and $\mu = -1$ (in the *background*) to 1 (in the *foreground*)

trajectory many times and constructing a histogram. Similar observations were made earlier numerically [9, 155, 156, 160] and T. Dittrich in [42].

7.4.3 Expectation Values

Suppose that a system is prepared at time $t = 0$ by specifying the density matrix $\rho(0)$, or equivalently the initial Wigner function $\rho_W(\boldsymbol{x}, 0)$. We let the system evolve for a discrete time t and then measure any observable $\hat{a}$ of interest. The expectation value of the observable is given by

$$\langle a(t) \rangle \equiv \text{tr}[\hat{a}\rho(t)] = \int d\boldsymbol{x}\, a_W(\boldsymbol{x})\rho_W(\boldsymbol{x}, t), \qquad (7.67)$$

where $a_W(\boldsymbol{x})$ is the Weyl symbol associated with the operator $\hat{a}$. The definition of a is analogous to the definition of ρ_W [42]. To lowest order in $1/J$, $a_W(\boldsymbol{x})$ equals the classical observable $a(\boldsymbol{x})$ that corresponds to $\hat{a}$, if the classical observable exists. Using (7.63), we immediately obtain

$$\langle a(t) \rangle = \int d\boldsymbol{x}\, a(\boldsymbol{x}) P_{\text{cl}}^t \rho_W(\boldsymbol{x}, 0), \qquad (7.68)$$

up to corrections of order $1/J$. Thus, quantum mechanical expectation values can be obtained from a knowledge of the classical propagator and the classical observable for any initial Wigner function that contains no structure on the scale of $1/J$. Equation (7.68) is a hybrid classical–quantum formula, since the initial Wigner function can be very *nonclassical*, e.g. it can contain regions where $\rho_W(\boldsymbol{x}, 0) < 0$.

But interesting cases also arise where the initial Wigner function *is* a classical phase space density, $\rho_W = \rho_{\text{cl}}$, or where the time t is sent to infinity, so that the invariant state is reached. In both cases the quantum mechanical expectation value is given by a purely classical formula. In particular, expectation values in the invariant state $\rho_W(\infty)$ are given by

$$\langle a \rangle_\infty = \int d\boldsymbol{x}\, a(\boldsymbol{x})\rho_{\text{cl}}(\boldsymbol{x}, \infty), \qquad (7.69)$$

since, up to corrections of order $1/J$, $P_{\text{cl}}\rho_W(\infty) = P_{\text{cl}}\rho_{\text{cl}}(\infty) = \rho_{\text{cl}}(\infty)$. This allows us to use highly developed classical-periodic orbit theories [14, 33] to evaluate $\langle a \rangle_\infty$ (see next section).

7.4.4 Correlation Functions

The discrete time correlation function $K(t_2, t_1)$ between two observables a and b with respect to an initial density matrix $\rho(0)$ is defined as [186]

$$K(t_2, t_1) = \langle b(t_2)a(t_1) \rangle_0 \equiv \text{tr}[bP^{t_2}aP^{t_1}\rho(0)]. \qquad (7.70)$$

7.4 Consequences

This function has typically a real and an imaginary part. The latter is connected to the Fourier transform with respect to frequency of a linear susceptibility (see, e.g., [97]). Here I show how the real part of $K(t_2, t_1)$ can be calculated semiclassically.

The starting point is the observation that in (7.70) $aP^{t_1}\rho(0) = a\rho(t_1)$ enters in the same way as the initial density matrix $\rho(0)$ enters in $\langle a(t)\rangle$, (cf. (7.67)). In fact, formally, $K(t_2, t_1)$ is nothing but the expectation value of b with respect to the "density matrix" $\rho'(t_1) \equiv a\rho(t_1)$ propagated t_2 times. Note that $\rho'(t_1)$ is not really a density matrix, in general not even a Hermitian operator. However, in the derivation of expectation values the special properties of a density matrix (in addition to Hermiticity, also positivity and a trace equal to unity) did not enter at any point. The only thing that did matter was that the density matrix had to have a smooth Wigner transform. This I shall suppose as well about the Wigner transform $\rho'_W(t_1)$ of $\rho'(t_1)$. Later on, we shall see that the assumption is self-consistent if the initial density matrix $\rho(0)$ has a smooth Wigner transform and the Weyl symbol a_W a smooth classical limit. So let us introduce a Wigner transform $\rho'_W(\boldsymbol{x}, t_1)$ in complete analogy to the definition (7.53) for any density matrix,

$$\rho'_W(\mu, \phi) = \frac{J^2}{\pi} \int d\eta\, e^{2iJ\eta\phi} \rho'(\mu, \eta, t_1)\,, \qquad (7.71)$$

and then use (7.68) to express the correlation function $K(t_2, t_1)$ as

$$K(t_2, t_1) = \int d\boldsymbol{x}\, d\boldsymbol{y}\, b_{cl}(\boldsymbol{y}) P_{cl}^{t_2} \rho'_W(\boldsymbol{x}, t_1)\,. \qquad (7.72)$$

We can write $\rho'(\mu, \eta, t_1) = \langle m+k|a\rho(t_1)|m-k\rangle$ in (7.71) as

$$\rho'(\mu, \eta, t_1) = J \int d\lambda \sum_{n=-\infty}^{\infty} \langle m+k|a|J\lambda\rangle\langle J\lambda|\rho(t_1)|m-k\rangle e^{iJ2\pi n\lambda}\,, \qquad (7.73)$$

where I have introduced a factor of unity with $l = J\lambda$ as a summation variable and then changed the sum to an integral over l by Poisson summation. In terms of the corresponding Weyl symbol and Wigner function, we have

$$\langle m+k|a|l\rangle = \frac{1}{2\pi} \int d\phi_1 e^{\left(-iJ(\mu+\eta-\lambda)\phi_1\right)} a_W\left(\frac{\mu+\eta+\lambda}{2}, \phi_1\right)\,,$$

$$\langle l|\rho(t_1)|m-k\rangle = \frac{1}{J} \int d\phi_2 e^{\left(-iJ(\lambda-\mu+\eta)\phi_2\right)} \rho_W\left(\frac{\lambda+\mu-\eta}{2}, \phi_2, t_1\right)\,.$$

If we insert the last two equations into (7.73) and the resulting equation into (7.71), we are led to

$$\rho'_W(\mu, \phi, t_1) = \frac{J^2}{2\pi^2} \sum_{n=-\infty}^{\infty} \int d\lambda\, d\eta\, d\phi_1\, d\phi_2\, \exp\left(iJH(\lambda, \eta, \phi_1, \phi_2)\right)$$

$$\times a_W\left(\frac{\mu+\eta+\lambda}{2}, \phi_1\right) \rho_W\left(\frac{\lambda+\mu-\eta}{2}, \phi_2, t_1\right)\,, \qquad (7.74)$$

with an exponent H given by

$$H(\lambda,\eta,\phi_1,\phi_2) = 2\eta\phi - (\mu+\eta-\lambda)\phi_1 + (\mu-\eta-\lambda)\phi_2 + 2\pi n\lambda. \quad (7.75)$$

Integration by the SPA leads to the saddle-point equations

$$\partial_\eta H = 2\phi - \phi_1 - \phi_2 = 0, \quad (7.76)$$
$$\partial_\lambda H = \phi_1 - \phi_2 + 2\pi n = 0, \quad (7.77)$$
$$\partial_{\phi_1} H = -(\mu+\eta-\lambda) = 0, \quad (7.78)$$
$$\partial_{\phi_2} H = \mu - \eta - \lambda = 0. \quad (7.79)$$

The second equation gives immediately $\phi_1 = \phi_2 \bmod 2\pi$, and if we restrict ϕ_1,ϕ_2 as before to an interval of 2π, we have $n=0$ and $\phi_1 = \phi_2 = \phi$ from (7.76). The last two equations give $\eta = 0$ and $\lambda = \mu$. The value of H at the saddle point is zero and one can easily check that the determinant of second derivatives gives 4. Putting all pieces of the SPA together, we obtain

$$\rho'_W(\boldsymbol{x},t_1) = a_W(\boldsymbol{x})\rho_W(\boldsymbol{x},t_1). \quad (7.80)$$

This means that $\rho'_W(\boldsymbol{x},t_1)$ is smooth if $a_W(\boldsymbol{x})$ and $\rho_W(\boldsymbol{x},t_1)$ are smooth. If we remember that to lowest order in $1/J$ the Weyl symbols a_W and b_W are just the classical observables a and b, we obtain from (7.72) the final result

$$K(t_2,t_1) = \int d\boldsymbol{x}\, b(f^{t_2}(\boldsymbol{x}))a(\boldsymbol{x})\rho_W(\boldsymbol{x},t_1). \quad (7.81)$$

So, semiclassically, the correlation function has the same structure as a classical correlation function with respect to a classical phase space density at time t_1, namely $\rho_{\rm cl}(\boldsymbol{x},t_1)$,

$$K_{\rm cl}(t_2,t_1) = \int d\boldsymbol{x}\, b(f^{t_2}(\boldsymbol{x}))a(\boldsymbol{x})\rho_{\rm cl}(\boldsymbol{x},t_1). \quad (7.82)$$

The only remainder of quantum mechanics is the Wigner function after t_1 steps. We have the same kind of hybrid classical–quantum formula as for expectation values. And as for expectation values, in the limit of large t_1 and with $t_2 - t_1 = t$ kept fixed, the quantum mechanical correlation function approaches its classical value, as $\rho_W(\boldsymbol{x},t_1)$ tends to the smeared-out classical invariant state $\rho_{\rm cl}(\boldsymbol{x},\infty)$. Nevertheless, as pointed out in the context of expectation values, $\rho_W(\boldsymbol{x},t_1)$ can describe very nonclassical states, for instance Schrödinger cat states [142].

It is also remarkable that the expression in (7.81) is always real. We can trace this back to the realness of ρ'_W in (7.80). Since $a\rho(t_1)$ is not necessarily Hermitian, there would seem to be no need for ρ'_W to be real. However, note that $a\rho(t_1)$ would be Hermitian, if a and $\rho(t_1)$ commuted. Since they do commute classically, the commutator must be of order $1/J$, and the imaginary part in $K(t_2,t_1)$ is therefore always at least one order in $1/J$ smaller than the real part.

If $t_1 \to \infty$ (with $t = t_2 - t_1$ fixed) or if $\rho(\boldsymbol{x},0)$ is chosen as the invariant density matrix so that $K(t+t_1,t_1)_{t_1\to\infty} \equiv K(t) = K_{\rm cl}(t) \equiv K_{\rm cl}(t+t_1,t_1)_{t_1\to\infty}$, we can use classical periodic-orbit theory to calculate the quantum mechanical correlation function [187]. The use of the theory is completely analogous to the case of expectation values and will be discussed in more detail in the next section.

7.5 Trace Formulae for Expectation Values and Correlation Functions

Cvitanović and Eckhardt [33] have generalized the periodic-orbit theory for the Frobenius–Perron propagator to other operators and have obtained a very precise tool for the calculation of classical expectation values in the invariant state. I shall briefly present this theory here, since it is a natural complement to the quantum-classical hybrid formulae encountered in the last section, which become entirely classical for initial Wigner functions representing the invariant state.

7.5.1 The General Strategy

We have defined classical expectation values $\langle a \rangle_\infty$ so far with respect to an invariant classical phase space density ρ_{cl}. We can also consider time averages,

$$\overline{a(x)} = \lim_{t\to\infty} \frac{1}{t} A(x,t), \qquad A(x,t) = \sum_{n=0}^{t-1} a(f_{\mathrm{cl}}^t(x)). \tag{7.83}$$

If the system is ergodic (I shall consider *only* ergodic, dissipative systems in the following), the two averages are the same except for starting points x of measure zero, like for example periodic points. The influence of such special points can be excluded by averaging the starting point over all accessible phase space;

$$\langle \overline{a(x)} \rangle = \frac{1}{\Omega} \int \mathrm{d}x\, \overline{a(x)}. \tag{7.84}$$

We then have $\langle a \rangle_\infty = \langle \overline{a(x)} \rangle$ for an ergodic system and can concentrate our efforts on calculating the latter quantity.

It is convenient to define a generating function

$$s(\gamma) = \lim_{t\to\infty} \frac{1}{t} \ln \langle e^{\gamma A(x,t)} \rangle, \tag{7.85}$$

as the expectation value sought is then easily expressed as

$$\langle \overline{a(x)} \rangle = \left.\frac{\partial s(\gamma)}{\partial \gamma}\right|_{\gamma=0}. \tag{7.86}$$

Fluctuations of a can be obtained from the second derivative if desired. I now show that $s(\gamma)$ can be extracted as the logarithm of the leading eigenvalue

112   7. Semiclassical Analysis of Dissipative Quantum Maps

of a generalized Frobenius–Perron propagator P_A^t defined by the phase space representation [14, 33]

$$P_A^t(\boldsymbol{y}, \boldsymbol{x}) = \delta(\boldsymbol{y} - \boldsymbol{f}_{\mathrm{cl}}^t(\boldsymbol{x})) e^{\gamma A(\boldsymbol{x}, t)}. \tag{7.87}$$

This operator acts on a phase space density $\rho(\boldsymbol{x})$ according to

$$(P_A^t \rho)(\boldsymbol{y}) = \int d\boldsymbol{x} P_A^t(\boldsymbol{y}, \boldsymbol{x}) \rho(\boldsymbol{x}). \tag{7.88}$$

The generalized propagator has a semigroup property. And, in analogy to the Frobenius–Perron operator (which is recovered for $\gamma = 0$), we assume that a spectral decomposition

$$P_A^t = \sum_n |n\rangle\langle n| e^{s_n(\gamma) t} \tag{7.89}$$

exists, with an ordering of the eigenvalues $\exp(s_n(\gamma))$ such that $\mathrm{Re}(s_0) > \mathrm{Re}(s_1) \geq \ldots$ As the system is supposed to be ergodic and dissipative, the leading eigenvalue $\exp(s_0)$ is, at $\gamma = 0$, equal to unity and nondegenerate.

One can easily convince oneself that

$$\langle e^{\gamma A(\boldsymbol{x}, t)} \rangle = \langle P_A^t 1 \rangle = \sum_n c_n e^{s_n(\gamma) t}, \tag{7.90}$$

where the form of the coefficients c_n is of no further concern. For if we now calculate $s(\gamma)$ from the definition (7.85) with the help of (7.90), the sum over n is dominated for large t entirely by $s_0(\gamma)$, and we obtain

$$s(\gamma) = \lim_{t \to \infty} \frac{1}{t} \ln \left(\sum_n c_n e^{s_n(\gamma) t} \right) = s_0(\gamma). \tag{7.91}$$

So *the generating function $s(\gamma)$ is the logarithm of the leading eigenvalue of the generalized Frobenius–Perron propagator P_A*.

The rest of the formalism aims at calculating this leading eigenvalue. It is most conveniently obtained from a spectral determinant,

$$F(z) = \det(1 - zP_A) = \exp\left(-\sum_{n=1}^{\infty} \frac{z^n}{n} \mathrm{tr}\, P_A^n\right), \tag{7.92}$$

which has roots at $z_\alpha(\gamma) = \exp(-s_\alpha(\gamma))$. Note that so far this is just a formal definition, which only becomes useful because an expansion of the exponential function on the right-hand side in powers of z leads to a polynomial with rapidly decaying coefficients. For the further exploitation of the formalism, two different strategies are possible.

7.5.2 Cycle Expansion

The traditional strategy is a so-called cycle expansion, briefly alluded to in Sect. 7.2.3. This means that the expansion of the exponential is performed after inserting the trace formula for P_A^t,

7.5 Trace Formulae for Expectation Values and Correlation Functions

$$\operatorname{tr} P_A^t = \sum_{\text{p.o.}} n_p \sum_r \frac{e^{r\gamma A_p} \delta_{t,n_p r}}{|\det \mathbf{1} - \mathbf{M}_p^r|}, \qquad (7.93)$$

where $A_p = A(\boldsymbol{x}_p, n_p)$ is the observable summed along the primitive periodic orbit p, with length n_p, starting at $\boldsymbol{x}_p$. Equation (7.93) is the result of a simple two-line calculation exploiting the properties of the delta function. By expanding the exponential, one combines different prime cycles systematically to obtain so-called pseudo-orbits or pseudo-cycles π. There are very many of them, since a pseudo-cycle is defined as a distinct nonrepeating combination $\{p_1, \ldots, p_k\}$ of prime cycles with a given total topological length $n_\pi = n_{p_1} + \ldots + n_{p_k}$.[1] The stabilities of a prime cycle are the (in the present context, two) stability eigenvalues, i.e. the eigenvalues of the Jacobian $\mathbf{M}_p$ connected with the map from the starting point of the prime cycle to the last point before the prime cycle closes. For dissipative maps the product of the two eigenvalues usually does not equal unity, so we need always to calculate both of them. In the following, Λ_p will denote the product of all expanding eigenvalues (i.e. with absolute value larger than unity) of a prime cycle p, and 1 if both are contracting. The stability product enters into the pseudo-cycle weight $t_\pi = (-1)^{k+1}/|\Lambda_{p_1} \ldots \Lambda_{p_k}|$, where k denotes the number of prime cycles involved. The values A_p of the observable summed along the prime cycles are combined to give a corresponding quantity for the pseudo-cycles as well, $A_\pi = A_{p_1} + \ldots + A_{p_k}$. With all this, the expectation value of the classical observable A in the invariant state, $\langle A \rangle_\infty$, is given by [14, 33]

$$\langle A \rangle_\infty = \frac{\sum_\pi A_\pi t_\pi}{\sum_\pi n_\pi t_\pi}. \qquad (7.94)$$

The big advantage of the cycle expansion is that the original sums over periodic orbits are truncated in a clever way, such that almost-compensating terms are grouped together. This leads to a rapid convergence of the leading eigenvalues. The starting point for the practical use of (7.94) is a list of prime cycles of the classical map, their stabilities, their topological lengths and the observable summed along the prime cycle. This information has to be calculated numerically.

Figure 7.14 shows a comparison of some exact quantum mechanical results for the observables J_z/J and J_y/J in the invariant state, compared to results from classical periodic-orbit theory (7.94), and results from straightforward classical evaluations. The latter were performed by iterating many randomly chosen initial phase space points and averaging over the trajectories generated. Whereas J_y fluctuates only slightly about the value of zero suggested by the symmetry of the problem, J_z/J decreases from 0 for zero dissipation to -1 for strong dissipation as the strange attractor shrinks more and more towards the south pole of the Bloch sphere [189]. The figure shows that even with rather short orbits ($n_\pi \leq 4$), the quantum mechanical result

---
[1] Instead of truncating the sums of periodic orbits at a given total topological length, also stability ordering has been considered [188].

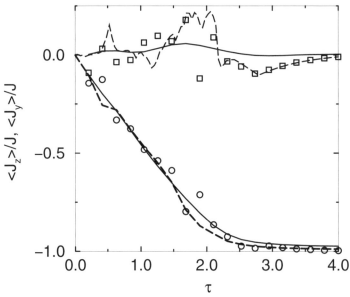

Fig. 7.14. Semiclassically calculated expectation values $\langle J_z \rangle/J$ and $\langle J_y \rangle/J$ as a function of dissipation τ for $k = 8.0$, $\beta = 2.0$ compared with a direct quantum mechanical evaluation ($j = 20$, *full lines*) and the classical expectation values (*dashed lines*). Close to $\tau = 0.5$ and $\tau = 2.0$ the agreement is somewhat spoiled by a nearby bifurcation

is reproduced very well for both observables. The agreement improves, as expected, with larger values of J and becomes almost perfect when comparing the classical simulation with (7.94).

A classical periodic-orbit theory for correlation functions was invented by Eckhardt and Grossmann [187]. It is completely analogous to the theory for the expectation values. In fact, the classical correlation function is nothing but the expectation value of $b(t)a(0)$ in the invariant state $\rho_{\text{cl}}(\infty)$, so that in (7.94) we just insert for A_p the variable $b(t)a(0)$ averaged along the prime cycle p. The practical evaluation of $K(t)$ via the periodic-orbit formula is, however, handicapped by the fact that to obtain $K(t)$ one must have prime cycles of at least length t. Finding all of these for large t is a difficult numerical problem, and is hindered additionally in our example by the fact that we do not have a symbolic dynamics for the dissipative kicked top. Nevertheless, Fig. 7.15 shows that at least the classical result and the real part of the quantum mechanical correlation function $\langle J_z(t) J_z(0) \rangle$ agree rather well.

7.5.3 Newton Formulae for Expectation Values

In spite of their success, cycle expansions are rather demanding from a numerical point of view since they involve the combinatorial problem of regrouping the prime cycles into nonrepeating pseudo-cycles. The effort involved in these

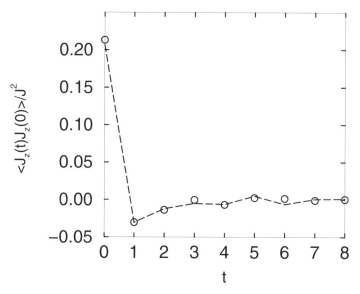

Fig. 7.15. Quantum mechanical correlation function of J_z/J ($j = 40$, *circles*) for $k = 8.0$, $\beta = 2.0$, $\tau = 1.0$ compared with direct classical evaluation (*dashed line*)

combinatorics increases exponentially with the maximum pseudo-cycle length n_π used.

On the other hand, we have seen in the context of the leading eigenvalues of P_{cl} that Newton's formulae originate from the very same expansion as the cycle expansion. They therefore already contain implicitly the whole combinatorics of all pseudo-cycles to any desired order. Now a much more efficient strategy becomes evident: instead of recombining periodic orbits after the expansion of the logarithm, with great numerical effort, into pseudo-cycles, we can just calculate the traces (7.93) numerically for a very small value $\gamma \ll 1$ from the very same cycle list, use Newton's formulae to obtain the coefficients of the characteristic polynomial (7.92), and then find its roots z_α. The roots close to unity can, in general, be easily obtained. Finally, we approximate the desired derivative numerically according to

$$\left.\frac{\partial s_0(\gamma)}{\partial \gamma}\right|_{\gamma=0} \simeq -\frac{\ln z_0(\gamma) - \ln z_0(0)}{\gamma}. \tag{7.95}$$

Note that in (7.94) it was assumed that $z_0(0) = 1$, which must hold if the exponential is truncated at very high powers. However, in reality the leading eigenvalue can be rather far from unity. Including the actual value of $z_0(0)$ in (7.95) is essential and can lead to results different from those of the cycle expansion.

Figure 7.16 shows a comparison of the two methods. For the same maximum length of the cycles included, both give results of comparable accuracy.

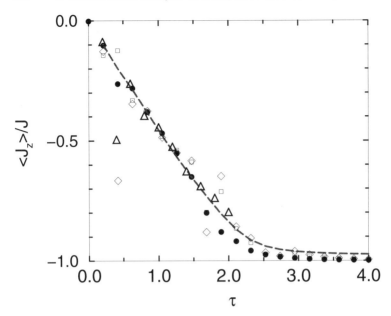

Fig. 7.16. Comparison of several methods for the calculation of the expectation value $\langle J_z \rangle / J$ for the dissipative kicked top with $k = 8.0$, $\beta = 2.0$. The classical result (*filled circles*) and the quantum mechanical result for $j = 20$ (*dashed line*) are shown, together with results from a cycle expansion including all terms up to $n_\pi = 4$ (*squares*), from Newton's formulae with $n_p \leq 4$ (*diamonds*), and from Newton's formulae with $n_p \leq 7$ (*triangles*, up to $\tau = 2$). Close to $\tau \simeq 0.5$ and 2.0, bifurcations spoil the semiclassical results (cycle expansion and Newton's formulae), but otherwise the quantum mechanical result is reproduced very well

However, the method using Newton's formulae has the advantage of enormous numerical simplification. A calculation for cycles up to a length $t = 7$ takes a few seconds, compared with many hours for the cycle expansion. The value of the parameter γ could be varied over a large region. Values ranging from 10^{-3} to 10^{-6} yielded the same results.

7.6 Summary

In this final chapter I have shown how a semiclassical propagator P can be obtained for dissipative quantum maps of the type introduced in Chap. 6. I have used the propagator to derive a trace formula for $\operatorname{tr} P^t$, which turned out to be the trace formula for the classical Frobenius–Perron operator, $\operatorname{tr} P_{\text{cl}}^t$. I have shown that the leading eigenvalues of P approach the Ruelle resonances of P_{cl} if the effective $\hbar$ in the system, namely the parameter $1/j$, goes to zero. The leading eigenvalues of P could be extracted from the trace formula by means of Newton's formulae, and we have seen how this procedure is limited numerically. In the last section I have used the same procedure to calculate quantum mechanical expectation values in the invariant state. I have derived the propagator for the Wigner function and have shown that the expectation values (and correlation functions) in the invariant state are given to lowest order in $1/j$ by classical formulae.

A. Saddle-Point Method for a Complex Function of Several Arguments

Let $\boldsymbol{x}$ represent M real variables $x_1, \ldots, x_M$, and let $F(\boldsymbol{x})$ and $G(\boldsymbol{x})$ be complex-valued functions of $\boldsymbol{x}$ with $\operatorname{Re} F \leq 0$ in the volume V in $\mathcal{R}^M$ over which we are going to integrate. Suppose that V contains a single nondegenerate stationary point $\boldsymbol{x}_0$ with $\partial_{x_i} F(\boldsymbol{x}_0) = 0$, $i = 1, \ldots, M$. Let us denote by $Q_{M \times M}$ the matrix of the negative second derivatives, i.e. $(Q_{M \times M})_{ik} = -\partial_{x_i} \partial_{x_k} F(\boldsymbol{x})$ taken at $\boldsymbol{x} = \boldsymbol{x}_0$. The condition of nondegeneracy means $\det \mathbf{Q}_{M \times M} \neq 0$. We then have, for $J \to \infty$ [190],

$$\int_V d^M \boldsymbol{x}\, e^{JF(\boldsymbol{x})} G(\boldsymbol{x})$$
$$= G(\boldsymbol{x}_0) \sqrt{\frac{(2\pi)^M}{J^M |\det Q_{M \times M}|}} e^{JF(\boldsymbol{x}_0) - (i/2) \operatorname{Ind} Q_{M \times M}} [1 + \mathcal{O}(1/J)]. \quad (A.1)$$

Here $\operatorname{Ind} Q_{M \times M}$ is the index of the complex quadratic form χ of M real variables,

$$\chi = \sum_{i,j=1}^{M} (Q_{M \times M})_{ij} x_i x_j. \quad (A.2)$$

The index is defined via the minors $D_k = \det ||(Q_{M \times M})_{ij}||$, $1 \leq i, j \leq k$ of $Q_{M \times M}$ as

$$\operatorname{Ind} Q_{M \times M} = \sum_{k=1}^{M} \arg \rho_k, \quad -\pi < \arg \rho_k \leq \pi, \quad (A.3)$$

$$\rho_1 = D_1 = (Q_{M \times M})_{11}, \quad \rho_k = \frac{D_k}{D_{k-1}}, \quad k = 2, \ldots, M. \quad (A.4)$$

The restriction $-\pi < \arg \rho_k \leq \pi$ on the phases fixes uniquely the overall phase of the saddle-point contribution. Without this restriction, $\operatorname{Ind} Q_{M \times M}$ would be defined only up to multiples of 2π, and this would lead to an overall phase ambiguity corresponding to the choice of sign for the square root in (A.1).

If any of the minors D_k is zero we add a term $-\epsilon(x_i - x_{0_i})^2$ to the function $F(\boldsymbol{x})$. Such an addition does not change the convergence of the integral (in the limit of $J \to \infty$), nor its value for $\epsilon \to 0$. However, such a small term may bring the D_k away from zero and therefore allow us to determine its phase.

The formula (A.1) has a structure familiar from the SPA of a complex function of a single variable. The term $-(i/2)\text{Ind}\, Q_{M\times M}$ leads to a phase analogous to Maslov's $\pm\pi/4$, but that phase can now take on any value between $-\pi/2$ and $\pi/2$.

B. The Determinant of a Tridiagonal, Periodically Continued Matrix

Let A be a $t \times t$ matrix with the structure

$$A = \begin{pmatrix} a_{11} & a_{12} & 0 & 0 & \cdots & 0 & a_{1t} \\ a_{21} & a_{22} & a_{23} & 0 & \cdots & 0 & 0 \\ 0 & \ddots & \ddots & \ddots & & & 0 \\ \vdots & & & & & & \vdots \\ a_{t1} & 0 & \cdots\cdots\cdots & & & a_{t,t-1} & a_{tt} \end{pmatrix}. \tag{B.1}$$

Then $\det A$ can be expressed in terms of traces of 2×2 matrices formed from the original matrix elements according to [163]

$$\det A = \operatorname{tr} \prod_{j=t}^{1} \begin{pmatrix} a_{jj} & -a_{j,j-1} a_{j-1,j} \\ 1 & 0 \end{pmatrix} + (-1)^{t+1} \operatorname{tr} \prod_{j=t}^{1} \begin{pmatrix} a_{j,j-1} & 0 \\ 0 & a_{j-1,j} \end{pmatrix}. \tag{B.2}$$

The proof of the formula is quite analogous to the solution of a Schrödinger equation for a one-dimensional tight-binding Hamiltonian with nearest-neighbor hopping by using transfer matrices [163]. The inverse order of the initial and final indices on the product symbol indicates that the matrix with the highest index j is on the left of the product. The formula should only be applied for $t \geq 3$.

C. Partial Classical Maps and Stability Matrices for the Dissipative Kicked Top

I collect together here the classical maps for the three components rotation, torsion and dissipation for the kicked top studied in this book, as well as their stability matrices in phase space coordinates. All maps will be written in the notation $(\mu, \phi) \longrightarrow (\nu, \psi)$, i.e. μ and ν stand for the initial and final momentum, and ϕ and ψ for the initial and final (azimuthal) coordinate. The latter is defined in the interval from $-\pi$ to π. The stability matrices will be arranged as

$$\mathbf{M} = \begin{pmatrix} \partial \psi / \partial \phi & \partial \nu / \partial \phi \\ \partial \psi / \partial \mu & \partial \nu / \partial \mu \end{pmatrix}. \tag{C.1}$$

C.1 Rotation by an Angle β About the y Axis

The map reads

$$\nu = \mu \cos \beta - \sqrt{1 - \mu^2} \sin \beta \cos \phi, \tag{C.2}$$

$$\psi = \left\{ \arcsin \left(\sqrt{\frac{1 - \mu^2}{1 - \nu^2}} \sin \phi \right) \theta(x) \right.$$
$$\left. + \left[\operatorname{sign}(\phi) \pi - \arcsin \left(\sqrt{\frac{1 - \mu^2}{1 - \nu^2}} \sin \phi \right) \theta(-x) \right] \right\} \bmod 2\pi, \tag{C.3}$$

$$x = \sqrt{1 - \mu^2} \cos \phi \cos \beta + \mu \sin \beta, \tag{C.4}$$

where x is the x component of the angular momentum after rotation, $\theta(x)$ is the Heaviside theta function and $\operatorname{sign}(x)$ denotes the sign function.

The stability matrix connected with this map is

$$\mathbf{M}_r = \begin{pmatrix} M_{r11} & M_{r12} \\ M_{r21} & M_{r22} \end{pmatrix} \tag{C.5}$$

where

$$M_{r11} = \sqrt{1 - \mu^2} \left(\frac{\cos \phi}{\sqrt{1 - \nu^2} \cos \psi} + \frac{\nu \sin \phi \tan \psi \sin \beta}{1 - \nu^2} \right),$$

$$M_{r12} = \sqrt{1 - \mu^2} \sin \phi \sin \beta,$$

$$M_{r21} = \frac{\nu \sin\psi(\sqrt{1-\mu^2}\cos\beta + \mu\cos\phi\sin\beta)}{\sqrt{1-\mu^2}(1-\nu^2)\cos\psi}$$

$$- \frac{\mu\sin\phi}{\sqrt{(1-\nu^2)(1-\mu^2)}\cos\psi},$$

$$M_{r22} = \cos\beta + \frac{\mu\cos\phi\sin\beta}{\sqrt{1-\mu^2}}.$$

C.2 Torsion About the z Axis

The map and stability matrix are given by

$$\nu = \mu, \tag{C.6}$$
$$\psi = (\phi + k\mu) \bmod 2\pi \tag{C.7}$$
$$\mathbf{M}_t = \begin{pmatrix} 1 & 0 \\ k & 1 \end{pmatrix}. \tag{C.8}$$

C.3 Dissipation

The dissipation conserves the angle ϕ, and the stability matrix is diagonal:

$$\nu = \frac{\mu - \tanh\tau}{1 - \mu\tanh\tau}, \tag{C.9}$$
$$\psi = \phi, \tag{C.10}$$
$$\mathbf{M}_d = \begin{pmatrix} 1 & 0 \\ 0 & (1-(\tanh\tau)^2)/(1-\mu\tanh\tau)^2 \end{pmatrix}. \tag{C.11}$$

The total stability matrix for the succession of rotation, torsion and dissipation is given by $\mathbf{M} = \mathbf{M}_d \mathbf{M}_t \mathbf{M}_r$.

References

1. H. Poincaré: *Les Méthodes Nouvelles de la Mécanique Céleste* (Gauthier-Villars, Paris, 1892)
2. A. Einstein: Verh. Dtsch. Phys. Ges. **19**, 82 (1917)
3. M.G. Gutzwiller: J. Math. Phys. **11**, 1791 (1970)
4. M.G. Gutzwiller: J. Math. Phys. **12**, 343 (1971)
5. O. Bohigas, M.J. Giannoni: in *Mathematical and Computational Methods in Nuclear Physics*, ed. by H. Araki, J. Ehlers, K. Hepp, R. Kippenhahn, H. Weidenmüller, J. Zittartz (Springer, Berlin, Heidelberg, 1984), Lecture Notes in Physics, Vol. 209
6. M.V. Berry: in *Chaotic Behavior of Deterministic Systems*, ed. by G. Iooss, R. Helleman, R. Stora (North-Holland, Amsterdam, 1981), Les Houches Session XXXVI
7. M.V. Berry: Proc. R. Soc. London A **400**, 229 (1985)
8. M.V. Berry, J.P. Keating: Proc. R. Soc. London A **437**, 151 (1992)
9. J. Iwaniszewski, P. Pepłowski: J. Phys. A: Math. Gen. **28**, 2183 (1995)
10. D. Wintgen, K. Richter, G. Tanner: Chaos **2**, 19 (1992)
11. D. Wintgen: Phys. Rev. Lett. **58**, 1589 (1987)
12. H. Hasegawa, M. Robnik, G. Wunner: Prog. Theor. Phys. Suppl. **98**, 198 (1989)
13. H. Schomerus, F. Haake: Phys. Rev. Lett. **79**, 1022 (1997)
14. P. Cvitanović, R. Artuso, R. Mainieri, G. Vatay: *Classical and Quantum Chaos* (Niels Bohr Institute, www.nbi.dk/ChaosBook/, Copenhagen, 2000)
15. P.A. Braun, D. Braun, F. Haake: Eur. Phys. J. D **3**, 1 (1998)
16. U. Smilansky: in *Mesoscopic Quantum Physics*, ed. by E. Akkermans, G. Montambaux, J.L. Pichard, J. Zinn-Justin (North-Holland, Amsterdam, 1994), Les Houches Session LXI
17. E. Ott: *Chaos in Dynamical Systems* (Cambridge University Press, Cambridge, 1993)
18. V.I. Arnold: AMS Transl. Ser. 2 **46**, 213 (1965)
19. R.M. May: Nature **261**, 459 (1976)
20. G.M. Zaslavsky: Phys. Lett. A **69**, 145 (1978)
21. D. Ruelle: *Chaotic Evolution and Strange Attractors* (Cambridge University Press, New York, 1989)
22. H. Furstenberg: Trans. Am. Math. Soc. **108**, 377 (1963)
23. V.I. Oseledets: Trans. Mosc. Math. Soc. **19**, 197 (1968)
24. A.N. Kolmogorov: Dokl. Akad. Nauk SSSR **119**, 861 (1958)
25. Y.G. Sinai: Russ. Math. Surveys **25**, 137 (1970)
26. R.C. Adler, A.C. Konheim, M.H. McAndrew: Trans. Am. Math. Soc. **114**, 309 (1965)
27. L. de Broglie: in *Rapport au Veme Congres de Physique Solvay* (Gauthier, Paris, 1930)

28. D. Bohm: Phys. Rev. **85**, 166 (1952)
29. J.S. Bell: *Speakable and Unspeakable in Quantum Mechanics* (Cambridge University Press, Cambridge, 1987)
30. G. Weihs, T. Jennewein, C. Simon, H. Weinfurter, A. Zeilinger: Phys. Rev. Lett. **81**, 5039 (1998)
31. M. Freyberger, M.K. Aravind, M.A. Horne, A. Shimony: Phys. Rev. A **53**, 1232 (1996)
32. P. Gaspard: *Chaos, Scattering and Statistical Mechanics* (Cambridge University Press, Cambridge, 1998)
33. P. Cvitanović, B. Eckhardt: J. Phys. A **24**, L237 (1991)
34. A.J. Lichtenberg, M.A. Liebermann: *Regular and Chaotic Dynamics*, 2nd edn. (Springer, New York, 1992)
35. D. Ruelle: Phys. Rev. Lett. **56**, 405 (1986)
36. D. Ruelle: J. Stat. Phys. **44**, 281 (1986)
37. D. Ruelle: J. Diff. Geom. **25**, 99 (1987)
38. D. Ruelle: Commun. Math. Phys. **125**, 239 (1989)
39. M. Policott: Invent. Math. **81**, 413 (1985)
40. M. Policott: Invent. Math. **85**, 147 (1986)
41. G. Floquet: Ann. de l'Ecole Norm. Sup. **12**, 47 (1883)
42. T. Dittrich: in *Quantum Transport and Dissipation*, ed. by T. Dittrich, P. Hänggi, G.L. Ingold, B. Kramer, G. Schön, W. Zwerger (Wiley–VCH, Weinheim, 1998)
43. D. Leonard, L.E. Reichl: J. Chem. Phys. **92**, 6004 (1990)
44. N.L. Balazs, A. Voros: Ann. Phys. (N.Y.) **190**, 1 (1989)
45. M. Saraceno: Ann. Phys. (N.Y.) **199**, 37 (1990)
46. A.M. Ozorio de Almeida: Ann. Phys. (N.Y.) **210**, 1 (1991)
47. B. Eckhardt, F. Haake: J. Phys. A: Math. Gen. **27**, 449 (1994)
48. P. Pakoński, A. Ostruszka, K. Życzkowski: Nonlinearity **12**, 269 (1999)
49. G. Casati, B.V. Chirikov, F.M. Izrailev, J. Ford: in *Stochastic Behavior in Classical and Quantum Hamiltonian Systems*, ed. by G. Casati, J. Ford (Springer, Berlin, Heidelberg, 1979), Vol. 93 of Lecture Notes in Physics
50. S. Fishman, D.R. Grempel, R.E. Prange: Phys. Rev. Lett. **49**, 509 (1982)
51. S. Fishman, D.R. Grempel, R.E. Prange: Phys. Rev. A **29**, 1639 (1984)
52. M. Feingold, S. Fishman, D.R. Grempel, R.E. Prange: Phys. Rev. B **31**, 6852 (1985)
53. D.L. Shepelyansky: Phys. Rev. Lett. **56**, 677 (1986)
54. G. Casati, J. Ford, I. Guarneri, F. Vivaldi: Phys. Rev. A **34**, 1413 (1986)
55. S.J. Chang, K.J. Shi: Phys. Rev. A **34**, 7 (1986)
56. A. Altland, M.R. Zirnbauer: Phys. Rev. Lett. **77**, 4536 (1996)
57. E.B. Bogomolny: Nonlinearity **5**, 805 (1992)
58. E.B. Bogomolny: Comments At. Mol. Phys **25**, 67 (1990)
59. F. Haake, M. Kuś, R. Scharf: in *Coherence, Cooperation, and Fluctuations*, ed. by F. Haake, L. Narducci, D. Walls (Cambridge University Press, Cambridge, 1986)
60. F. Haake: *Quantum Signatures of Chaos* (Springer, Berlin, 1991)
61. F. Waldner, D.R. Barberis, H. Yamazaki: Phys. Rev. A **31**, 420 (1985)
62. G.S. Agarwal, R.R. Puri, R.P. Singh: Phys. Rev. A **56**, 2249 (1997)
63. A. Peres: in *Quantum Chaos*, ed. by H.A. Cerdeira, R. Ramaswamy, M.C. Gutzwiller, G. Casati (World Scientific, Singapore, 1991)
64. A. Peres: *Quantum Theory: Concepts and Methods* (Kluwer Academic Publishers, Dordrecht, 1993)
65. R. Schack, C.M. Caves: Phys. Rev. Lett. **69**, 3413 (1992)
66. R. Schack, G. Ariano, C.M. Caves: Phys. Rev. E **50**, 972 (1994)

67. P.A. Miller, S. Sarkar: Phys. Rev. E **60**, 1542 (1999)
68. P.A. Miller, S. Sarkar: Nonlinearity **12**, 419 (1999)
69. O. Bohigas, M.J. Giannoni, C. Schmit: Phys. Rev. Lett. **52**, 1 (1984)
70. E.P. Wigner: Proc. Cambridge Philos. Soc. **47**, 790 (1951)
71. F.J. Dyson: J. Math. Phys. **3**, 140 (1962)
72. T.A. Brody, J. Flores, J.B. French, P.A. Mello, A. Pandey, S.M. Wong: Rev. Mod. Phys. **53**, 385 (1981)
73. M.L. Mehta: *Random Matrices*, 2nd edn. (Academic Press, New York, 1991)
74. L.E. Reichl: *The Transition to Chaos* (Springer, New York, 1992)
75. T. Guhr, A. Müller-Gröling, H.A. Weidenmüller: Phys. Rep. **299**, 190 (1998)
76. N. Rosenzweig, C.E. Porter: Phys. Rev. **120**, 1698 (1960)
77. M.V. Berry, M. Robnik: J. Phys. A: Math. Gen. **17**, 2413 (1984)
78. A. Altland, M.R. Zirnbauer: Phys. Rev. B **55**, 1142 (1997)
79. B.I. Shklovskii, B. Shapiro, B.R. Sears, P. Lambrianides, H.B. Shore: Phys.Rev.B **47**, 11 (1993)
80. E. Hofstetter, M. Schreiber: Phys. Rev. B **48**, 16 979 (1993)
81. D. Braun, G. Montambaux, M. Pascaud: Phys. Rev. Lett. **81**, 1062 (1998)
82. M. Barth, U. Kuhl, H.J. Stöckmann: Ann. Phys. (Berlin) **8**, 733 (1999)
83. P. Bertelsen, C. Ellegaard, T. Guhr, M. Oxborrow, K. Schaadt: Phys. Rev. Lett. **83**, 2171 (1999)
84. L.E. Reichl, Z.Y. Chen, M.M. Millonas: Phys. Rev. Lett. **63**, 2013 (1989)
85. R.P. Feynman, A.R. Hibbs: *Quantum Mechanics and Path Integrals* (McGraw-Hill, New York, 1965)
86. M.S. Marinov: Phys. Rep. **60**, 1 (1980)
87. M.G. Gutzwiller: *Chaos in Classical and Quantum Mechanics* (Springer, New York, 1991)
88. F. Haake: *Quantum Signatures of Chaos*, 2nd edn. (Springer, Berlin, Heidelberg, 2000)
89. J.H. Van Vleck: Proc. Natl. Acad. Sci. USA **14**, 178 (1928)
90. S.C. Creagh, J.M. Robbins, R.G. Littlejohn: Phys. Rev. A **42**, 1907 (1990)
91. M. Tabor: Physica D **6**, 195 (1983)
92. P.A. Braun, P. Gerwinski, F. Haake, H. Schomerus: Z. Phys. B **100**, 115 (1996)
93. P.A. Braun: Opt. Spektrosk. (USSR) **66**, 32 (1989)
94. P.A. Braun: Rev. Mod. Phys **65**, 115 (1993)
95. R.G. Littlejohn, J.M. Robbins: Phys.Rev.A **36**, 2953 (1987)
96. J.M. Robbins, R.G. Littlejohn: Phys. Rev. Lett. **58**, 1388 (1987)
97. U. Weiss: *Quantum Dissipative Systems* (World Scientific, Singapore, 1993)
98. R.P. Feynman, F.C. Vernon Jr.: Ann. Phys. (N.Y.) **24**, 118 (1963)
99. P. Ullersma: Physica (Utrecht) **32**, 215 (1966)
100. A.O. Calderia, A.J. Leggett: Phys. Rev. Lett. **46**, 211 (1981)
101. A.O. Calderia, A.J. Leggett: Ann. Phys. (N.Y.) **149**, 374 (1983)
102. H. Grabert, P. Schramm, G.L. Ingold: Phys. Rep. **168**, 115 (1988)
103. H. Grabert, H.R. Schober: in *Hydrogen in Metals III*, ed. by H. Wipf (Springer, Berlin, Heidelberg, 1997), Vol. 73 of Topics in Applied Physics
104. M.P.A. Fisher, W. Zwerger: Phys. Rev. B **32**, 6190 (1985)
105. D. Braun, U. Weiss: Physica B **202**, 264 (1994)
106. D. Cohen: Phys. Rev. Lett. **82**, 4951 (1999)
107. R. Bonifacio, P. Schwendiman, F. Haake: Phys. Rev. A **4**, 302 (1971)
108. M. Gross, C. Fabre, P. Pillet, S. Haroche: Phys. Rev. Lett. **36**, 1035 (1976)
109. G.S. Agarwal, A.C. Brown, L.M. Narducci, G. Vetri: Phys. Rev. A **15**, 1613 (1977)
110. M. Gross, S. Haroche: Phys. Rep. **93**, 301 (1982)

References

111. M.G. Benedict, A.M. Ermolaev, V.A. Malyshev, I.V. Sokolov, E.D. Trifonov: *Superradiance: Multiatomic Coherent Emission* (Institute of Physics Publishing, Bristol, 1996)
112. E.T. Jaynes, F.W. Cummings: Proc. IEEE **51**, 89 (1963)
113. G. Lindblad: Math. Phys. **48**, 119 (1976)
114. M. Gross, P. Goy, C. Fabre, S. Haroche, J.M. Raimond: Phys. Rev. Lett. **43**, 343 (1979)
115. N. Skribanowitz, I.P. Herman, J.C. MacGillivray, M.S. Feld: Phys. Rev. Lett. **30**, 309 (1973)
116. R. Bonifacio, P. Schwendiman, F. Haake: Phys. Rev. A **4**, 854 (1971)
117. P.A. Braun, D. Braun, F. Haake, J. Weber: Eur. Phys. J. D **2**, 165 (1998)
118. V.P. Maslov, V.E. Nazaikinskii: *Asymptotics of Operator and Pseudo-Differential Equations* (Consultants Bureau, New York, 1988)
119. K. Kitahara: Adv. Chem. Phys. **29**, 85 (1975)
120. E. Schrödinger: Die Naturwissenschaften **48**, 52 (1935)
121. W.H. Zurek: Phys. Rev. D **24**, 1516 (1981)
122. W.H. Zurek: Phys. Rev. D **26**, 1862 (1982)
123. E. Joos, H.D. Zeh: Z. Phys. B **59**, 223 (1985)
124. D.F. Walls, G.J. Milburn: Phys.Rev.A **31**, 2403 (1985)
125. W.H. Zurek: Phys. Today **44**(10), 36 (1991)
126. W.H. Zurek: Progr. Theor. Phys. **89**, 281 (1993)
127. W.H. Zurek, J.P. Paz: Phys. Rev. Lett **72**, 2508 (1994)
128. B.M. Garraway, P.L. Knight: Phys. Rev. A **50**, 2548 (1994)
129. S. Haroche: Phys. Today **51**(7), 36 (1998)
130. S. Habib, K. Shizume, W.H. Zurek: Phys. Rev. Lett. **80**, 4361 (1998)
131. J.P. Paz, W.H. Zurek: Phys. Rev. Lett. **82**, 5181 (1999)
132. D.A. Lidar, I.L. Chuang, K.B. Whaley: Phys. Rev. Lett. **81**, 2594 (1998)
133. L.M. Duan, G.C. Guo: Phys. Rev. A **57**, 2399 (1998)
134. A. Beige, D. Braun, P.L. Knight: Phys. Rev. Lett. **85**, 1762 (2000)
135. E.B. Wilson: J. Chem. Phys. **3**, 276 (1935)
136. H.C. Longuet-Higgins: Mol. Phys. **7**, 445 (1963)
137. J.H. Freed: J. Chem. Phys. **43**, 1710 (1965)
138. K.H. Stevens: J. Phys. C **16**, 5765 (1983)
139. W. Häusler, A. Hüller: Z. Phys. B **59**, 177 (1985)
140. L. Baetz: (1986), "Rechnersimulationen zur Temperaturabhängigkiet des Rotationstunnelns in Molekülkristallen", Ph.D. thesis, Universität Erlangen
141. A. Würger: Z. Phys. B **76**, 65 (1989)
142. D. Braun, P.A. Braun, F. Haake: Opt. Comm. **179**, 411 (1999)
143. J.F. Poyatos, J.J. Cirac, P. Zoller: Phys.Rev.Lett. **77**, 4728 (1996)
144. R.L. de Matos Filho, W. Vogel: Phys. Rev. Lett. **76**, 608 (1996)
145. F. Haake, R. Glauber: Phys. Rev. A **5**, 1457 (1972)
146. F.T. Arecchi, E. Courtens, R. Gilmore, H. Thomas: Phys. Rev. A **6**, 2211 (1972)
147. U. Fano: Rev. Mod. Phys. **29**, 74 (1957)
148. D.T. Smithey, M. Beck, M.G. Raymer, A. Faridani: Phys. Rev. Lett. **70**, 1244 (1993)
149. C. Kurtsiefer, T. Pfau, J. Mlynek: Nature **386**, 150 (1997)
150. D.S. Krahmer, U. Leonhardt: J. Phys. A: Math. Gen **30**, 4783 (1997)
151. G.M. D'Ariano: in *Quantum Communication, Computing, and Measurement*, ed. by P. Kumar, G. Ariano, O. Hirota (Plenum, New York, 1999)
152. M. Brune, E. Hagley, J. Dreyer, X. Maître, A. Maali, C. Wunderlich, J. Raimond, S.Haroche: Phys. Rev. Lett. **77**, 4887 (1996)
153. R.J. Cook: Phys. Scr. **T21**, 49 (1988)

154. R. Graham, T. Tél: Z. Phys. B **62**, 515 (1985)
155. T. Dittrich, R. Graham: Z. Phys. B **62**, 515 (1986)
156. T. Dittrich, R. Graham: Ann. Phys. (N.Y.) **200**, 363 (1990)
157. D. Cohen: J. Phys. A **31**, 8199 (1998)
158. P.A. Miller, S. Sarkar: Phys. Rev. E **58**, 4217 (1998)
159. T. Dittrich, R. Graham: Z. Phys. B **93**, 259 (1985)
160. P. Pepłowski, S.T. Dembiński: Z. Phys. B **83**, 453 (1991)
161. R. Grobe, F. Haake: Phys. Rev. Lett **62**, 2889 (1989)
162. R. Grobe: (1989), "Quantenchaos in einem dissipativen Spinsystem", Ph.D. thesis, Universität–GHS Essen
163. D. Braun, P. Braun, F. Haake: Physica D **131**, 265 (1999)
164. J. Ginibre: J. Math. Phys **6**, 440 (1965)
165. H.J. Sommers, A. Crisanti, H. Sompolinsky, Y. Stein: Phys. Rev. Lett. **60**, 1895 (1988)
166. N. Hatano, D.R. Nelson: Phys. Rev. Lett. **77**, 570 (1996)
167. M.A. Stephanov: Phys. Rev. Lett. **76**, 4472 (1996)
168. R.A. Janik, M.A. Nowak, G. Papp, I. Zahed: Phys. Rev. Lett. **77**, 4816 (1996)
169. Y.V. Fyodorov, B.A. Khoruzhenko, H.J. Sommers: Phys. Lett. A **226**, 46 (1997)
170. N. Hatano, D.R. Nelson: Phys. Rev. B **58**, 8384 (1998)
171. J.T. Chalker, B. Mehlig: Phys. Rev. Lett. **81**, 3367 (1998)
172. B. Mehlig, J.T. Chalker: J. Math. Phys. **41**, 3233 (2000)
173. F.R. Gantmacher: *Matrizentheorie* (Springer, Berlin, Heidelberg, 1986)
174. F. Christiansen, G. Paladin, H.H. Rugh: Phys. Rev. Lett. **17**, 2087 (1990)
175. E.R. Hanssen: *A Table of Series and Products* (Prentice Hall International, London, Sydney, Toronto, New Dehli, Tokyo, 1975)
176. J.H. Hannay, A.M. Ozorio de Almeida: J. Phys. A **17**, 3420 (1984)
177. J.F. Schipper: Phys. Rev. **184**, 1283 (1969)
178. D. Giulini, E. Joos, C. Kiefer, J. Kupsch, I.O. Stamtescu, H. Zeh: *Decoherence and the Appearance of a Classical World in Quantum Theory* (Springer, Berlin, Heidelberg, 1996)
179. E.P. Wigner: Phys. Rev. **40**, 1039 (1932)
180. M. Hillery, R.F. O'Connell, M.O. Scully, E.P. Wigner: Phys. Rep. **106**, 121 (1984)
181. R. Gilmore: Phys. Rev. A **12**, 1019 (1975)
182. G.S. Agarwal: Phys. Rev. A **24**, 2889 (1981)
183. R. Gilmore: in *The Physics of Phase Space: Nonlinear Dynamics and Chaos, Geometric Quantization and Wigner Function*, ed. by Y. Kim (Springer, Berlin, Heidelberg, 1987), proceedings of the 1. Internat. Conference on the Physics of Phase Space, held at the Univ. of Maryland
184. J.P. Dowling, G.S. Agarwal, W.P. Schleich: Phys. Rev. A **49**, 4101 (1994)
185. U. Leonhardt: Phys. Rev. A **53**, 2998 (1996)
186. F. Haake: *Statistical Treatment of Open Systems by Generalized Master Equations*, Vol. 66 of Springer Tracts in Modern Physics (Springer, Berlin, Heidelberg, 1973)
187. B. Eckhardt, S. Grossmann: Phys. Rev. E **50**, 4571 (1994)
188. P. Dahlqvist, G. Russberg: J. Phys. A **24**, 4763 (1991)
189. D. Braun: Chaos **9**, 730 (1999)
190. A.G. Prudkovsky: Zh. Vych. Mat. Mat. Fiz. [Journal of Computational Mathematics and Mathematical Physics] **14**, 299 (1974)

Index

angular-momentum coherent state   47, 55–60, 101, 105

bifurcation   68, 69, 85, 86, 88, 114, 116
box-counting dimension   16, 68

chaos   8, 9, 21, 24, 33, 63
– classical   1, 7–10, 18, 21, 22, 25
– quantum   2, 9, 21, 22, 24, 26, 27, 29, 63, 65
coupling agent   54, 56, 60
cycle expansion   93, 112–116

decoherence   3–5, 31, 37, 51–55, 60–63, 70, 83, 100
– accelerated   52–54, 57, 58
– in superradiance   55–59
decoherence-free subspace (DFS)   5, 53–55, 60–62
diagonal approximation   2, 3, 83, 84
Dyson's circular ensembles   25, 26

Ehrenfest time   2, 4, 84
ensemble description   7, 11, 12, 19, 26
ergodic measure   9, 15, 16, 64, 106
error generator   54

fixed point   3, 14, 15, 18, 66, 67, 85, 87, 88, 92
Frobenius–Perron propagator   3–5, 7, 11, 12, 17, 19, 64, 92, 101, 104, 111, 112, 116

generating properties of action   28, 79, 80
Ginibre's ensemble   70, 71, 73, 97

Heisenberg time   2, 4, 84

integrable dynamics   1, 8, 10, 11, 24–26, 66, 67, 69–71, 73, 83, 87

kicked top
– dissipative   5, 13, 16, 63, 65–67, 73, 85, 114, 116
– unitary   22, 25, 27, 36, 37

leading eigenvalues   5, 88, 91–96, 112, 115, 116
Lyapunov exponent   2, 9, 10, 18, 25, 67–69

map
– baker   8, 22
– cat   8
– classical   5, 7, 12, 21, 22, 29, 64, 66, 80, 113, 123
– Hamiltonian   12, 14, 15, 17, 29
– Henon   8
– logistic   8
– Poincaré   7, 8, 22
– quantum   4, 5, 14, 21, 22, 27, 29, 31, 63, 65, 75, 83
– – dissipative   5, 7, 18, 19, 22, 29, 49, 63, 65, 68, 90, 95, 97, 100, 116
– – unitary   21, 22, 28, 29, 36, 45, 68, 75
– sine circle   8
– standard   8, 22
– tent   8
Markovian approximation   33, 36, 37, 40, 49, 54
Maslov index   4, 28, 120
mixing   17
monodromy matrix   *see* stability matrix
Morse index   27, 28

Newton's formulae   29, 88, 91–93, 95, 114–116

periodic orbit   1, 3, 4, 14, 28, 80, 83, 84, 108, 111, 113–115
point attractor   15, 67, 86, 87

point repeller   3, 15, 66, 68, 69, 87, 89
pointer states   53, 56
Poisson summation   75, 76, 78, 102, 106
prime cycle   14, 113, 114
pseudo-orbit   93, 113

quantum computer   5, 54, 55, 62
quantum reservoir engineering   55

random-matrix conjecture   2, 25, 29
random-matrix theory (RMT)   2, 25, 95, 97
Ruelle resonance   4, 18, 84, 94–96, 116

saddle point approximation   2, 75, 77, 79–82, 86, 102–104, 110, 119
saddle-point approximation   75, 76
Schrödinger cat   4, 51, 55–59, 105, 110, 121
semiclassical approximation   2, 4, 33, 73
sensitivity
– of eigenvalues to changes of traces   89
– to changes of control parameters   24
– to initial conditions   1, 9–11, 24
spectral correlations   2, 97, 99
spectral density   1, 28
stability matrix   9, 13, 24, 27, 28, 82, 83, 85, 113, 123, 124

strange attractor   3, 13, 16, 66–68, 70, 73, 86, 87, 107, 113
superradiance
– classical behavior   36
– Hamiltonian dynamics   41, 42
– modeling   34, 36
– physics of   33, 34
– propagation of coherences   45
– semiclassical propagator   40
– short-time propagator   37
surface of section   7, 8, 15
symmetry
– for RMT ensembles   2, 25, 26, 64
– in coupling to environment   4, 35, 53, 54, 56, 58
– in coupling to environment   54

trace formula   1–5, 18, 21, 28, 29, 78, 83, 85, 86, 95, 106, 111, 112, 116

Van Vleck propagator   21, 27–29, 40, 44, 45, 47, 49

Wigner function   100–102, 104–109, 111, 117
Wigner surmise   26, 70
WKB approximation   3, 40–43, 47–49

zeta function   18, 92, 93, 95

Springer Tracts in Modern Physics

155 **High-Temperature-Superconductor Thin Films at Microwave Frequencies**
By M. Hein 1999. 134 figs. XIV, 395 pages

156 **Growth Processes and Surface Phase Equilibria in Molecular Beam Epitaxy**
By N.N. Ledentsov 1999. 17 figs. VIII, 84 pages

157 **Deposition of Diamond-Like Superhard Materials**
By W. Kulisch 1999. 60 figs. X, 191 pages

158 **Nonlinear Optics of Random Media**
Fractal Composites and Metal-Dielectric Films
By V.M. Shalaev 2000. 51 figs. XII, 158 pages

159 **Magnetic Dichroism in Core-Level Photoemission**
By K. Starke 2000. 64 figs. X, 136 pages

160 **Physics with Tau Leptons**
By A. Stahl 2000. 236 figs. VIII, 315 pages

161 **Semiclassical Theory of Mesoscopic Quantum Systems**
By K. Richter 2000. 50 figs. IX, 221 pages

162 **Electroweak Precision Tests at LEP**
By W. Hollik and G. Duckeck 2000. 60 figs. VIII, 161 pages

163 **Symmetries in Intermediate and High Energy Physics**
Ed. by A. Faessler, T.S. Kosmas, and G.K. Leontaris 2000. 96 figs. XVI, 316 pages

164 **Pattern Formation in Granular Materials**
By G.H. Ristow 2000. 83 figs. XIII, 161 pages

165 **Path Integral Quantization and Stochastic Quantization**
By M. Masujima 2000. 0 figs. XII, 282 pages

166 **Probing the Quantum Vacuum**
Pertubative Effective Action Approach in Quantum Electrodynamics and its Application
By W. Dittrich and H. Gies 2000. 16 figs. XI, 241 pages

167 **Photoelectric Properties and Applications of Low-Mobility Semiconductors**
By R. Könenkamp 2000. 57 figs. VIII, 100 pages

168 **Deep Inelastic Positron-Proton Scattering in the High-Momentum-Transfer Regime of HERA**
By U.F. Katz 2000. 96 figs. VIII, 237 pages

169 **Semiconductor Cavity Quantum Electrodynamics**
By Y. Yamamoto, T. Tassone, H. Cao 2000. 67 figs. VIII, 154 pages

170 **d–d Excitations in Transition-Metal Oxides**
A Spin-Polarized Electron Energy-Loss Spectroscopy (SPEELS) Study
By B. Fromme 2001. 53 figs. XII, 143 pages

171 **High-T_c Superconductors for Magnet and Energy Technology**
By B. R. Lehndorff 2001. 139 figs. XII, 209 pages

172 **Dissipative Quantum Chaos and Decoherence**
By D. Braun 2001. 22 figs. XI, 132 pages

Location: http://www.springer.de/phys/

You are one **click** *away from a* **world of physics** *information!*

Come and visit Springer's
Physics Online Library

Books
- Search the Springer website catalogue
- Subscribe to our free alerting service for new books
- Look through the book series profiles

You want to order?   Email to: orders@springer.de

Journals
- Get abstracts, ToC´s free of charge to everyone
- Use our powerful search engine LINK Search
- Subscribe to our free alerting service LINK *Alert*
- Read full-text articles (available only to subscribers of the paper version of a journal)

You want to subscribe?   Email to: subscriptions@springer.de

Electronic Media
- Get more information on our software and CD-ROMs

You have a question on
an electronic product?   Email to: helpdesk-em@springer.de

• Bookmark now:

http://www.springer.de/phys/

 Springer

Springer · Customer Service
Haberstr. 7 · D-69126 Heidelberg, Germany
Tel: +49 6221 345 200 · Fax: +49 6221 300186
d&p · 6437a/MNT/SF · Gha.

Printing: Mercedes-Druck, Berlin
Binding: Stürtz AG, Würzburg